Hybridisation as a conceptual framework for analysing / tracing the direction of change. But it does not offer an explanation for the drivers of change which require a multi-paradigm perspective.

Papers.

1. Oaklee - Competing Institutional Logics
2. Newington - Helm - Defender - Prospector.

DSD funding reduced from 70% to now 45%

Impact of hybridisation + competing institutional logics on corporate governance.

- Less responsive to tenant accountability - so tenant section needs to be sufficient.

Competing logics + impact on governance

(Professionalised smaller boards/Remuneration
(Tenant representations
(Accountability -

mergers / growth
Group Structures.

Hybridising Housing Organisations

Social and affordable housing has long been delivered through mixed economy mechanisms, but there has been little focus in housing studies on what this means for housing organisations themselves. This book presents recent international research applying concepts of social enterprise and hybridity to illuminate organisational behaviour in the housing sector. It addresses critiques of the explanatory value of these concepts by exploring their underlying meanings and their application to diverse case studies worldwide. The concepts are found to be most useful where they inform dynamic analysis of hybridisation and identify underlying change mechanisms, rather than simply providing static descriptions of hybridity. Various chapters in the book show how analysis can be enriched by drawing on institutional theory to develop concepts such as competing institutional logics, trade-offs between social and commercial goals and resource transfers. The book also looks at policy as a driver for hybridisation and to the regulatory challenges for policy systems that have come to rely on hybrid forms of delivery. A research agenda is proposed building on these conceptual frameworks to develop systematic approaches to data collection and analysis to enable clearer and more consistent meanings to emerge.

This book was published as a special issue of *Housing Studies* and includes an additional chapter on the same theme from an earlier issue of *Housing Studies*.

David Mullins is Professor of Housing Policy, Third Sector Research Centre and Housing and Communities Research Group at the University of Birmingham. His research interests include housing governance, management and regulation, homelessness, third sector, hybrid organisations and social enterprises and public services. He is on the Coordination Committee of the European Network for Social Housing.

Darinka Czischke is the Director of the Building and Social Housing Foundation and doctoral candidate at the Faculty of Architecture, Delft University of Technology. She was Director of the CECODHAS European Social Housing Observatory from 2005–2010. Her research interests include social housing, social enterprise, social innovation and socio-spatial integration.

Gerard van Bortel is a Lecturer and Researcher in Housing Management at the Faculty of Architecture and the Built Environment, Delft University of Technology.

Hybridising Housing Organisations

Meanings, Concepts and Processes of Social Enterprise in Housing

Edited by
David Mullins, Darinka Czischke and Gerard van Bortel

LONDON AND NEW YORK

First published 2014
by Routledge
2 Park Square, Milton Park, Abingdon, Oxon, OX14 4RN

Simultaneously published in the USA and Canada
by Routledge
711 Third Avenue, New York, NY 10017

Routledge is an imprint of the Taylor & Francis Group, an informa business

British Library Cataloguing in Publication Data
A catalogue record for this book is available from the British Library

ISBN 13: 978-0-415-70230-0

Typeset in Times New Roman
by Taylor & Francis Books

Publisher's Note
The publisher accepts responsibility for any inconsistencies that may have arisen during the conversion of this book from journal articles to book chapters, namely the possible inclusion of journal terminology.

Disclaimer
Every effort has been made to contact copyright holders for their permission to reprint material in this book. The publishers would be grateful to hear from any copyright holder who is not here acknowledged and will undertake to rectify any errors or omissions in future editions of this book.

Contents

Citation Information

The majority of chapters in this book were originally published in *Housing Studies*, volume 27, issue 4 (June 2012). When citing this material, please use the original page numbering for each article, as follows:

Chapter 1
Editorial: Exploring the Meaning of Hybridity and Social Enterprise in Housing Organisations
David Mullins, Darinka Czischke & Gerard van Bortel
Housing Studies, volume 27, issue 4 (June 2012)
pp. 405–417

Chapter 2
Conceptualising Social Enterprise in Housing Organisations
Darinka Czischke, Vincent Gruis & David Mullins
Housing Studies, volume 27, issue 4 (June 2012)
pp. 418–437

Chapter 3
The Quadruple Bottom Line and Nonprofit Housing Organizations in the United States
Rachel G. Bratt
Housing Studies, volume 27, issue 4 (June 2012)
pp. 438–456

Chapter 4
Entrenched Hybridity in Public Housing Agencies in the USA
Mai Thi Nguyen, William M. Rohe & Spencer Morris Cowan
Housing Studies, volume 27, issue 4 (June 2012)
pp. 457–475

Chapter 5
Let a Hundred Flowers Bloom: Innovation and Diversity in Australian Not-for-Profit Housing Organisations
Tony Gilmour & Vivienne Milligan
Housing Studies, volume 27, issue 4 (June 2012)
pp. 476–494

Chapter 6

Expansion, Diversification, and Hybridization in
Korean Public Housing
Hyunjeong Lee & Richard Ronald
Housing Studies, volume 27, issue 4 (June 2012)
pp. 495–513

Chapter 7

Negotiating Tensions: How Do Social
Enterprises in the Homelessness Field Balance
Social and Commercial Considerations?
Simon Teasdale
Housing Studies, volume 27, issue 4 (June 2012)
pp. 514–532

Chapter 8

Hybridity Enacted in a Large English Housing Association:
A Tale of Strategy, Culture and Community Investment
Halima Sacranie
Housing Studies, volume 27, issue 4 (June 2012)
pp. 533–552

Chapter 9

Magical or Monstrous? Hybridity in Social Housing
Governance
Anita Blessing
Housing Studies, volume 27, issue 2 (March 2012)
pp. 189–207

Please direct any queries you may have about the citations to
clsuk.permissions@cengage.com

INTRODUCTION

Exploring the Meaning of Hybridity and Social Enterprise in Housing Organisations*

DAVID MULLINS**, DARINKA CZISCHKE[†] & GERARD VAN BORTEL[‡]

**Third Sector Research Centre, University of Birmingham, [†]Faculty of Architecture, Delft University of Technology, [‡]OTB, Delft University of Technology

ABSTRACT   *While social housing has long been delivered through mixed economy mechanisms, there has been little focus in housing studies on what this means for housing organisations. This paper reviews recent international work applying concepts of social enterprise and hybridity to illuminate organisational behaviour. It addresses critiques of the explanatory value of these concepts by exploring their underlying meanings and their application to diverse case studies worldwide. The concepts are found to be most useful where they inform dynamic analysis of hybridisation and identify underlying change mechanisms, rather than simply providing static descriptions of hybridity. Analysis can be enriched by drawing on institutional theory to develop concepts such as competing organisational logics, trade-offs between social and commercial goals and resource transfers. The paper looks at policy as a driver for hybridisation and to the regulatory challenges for policy systems that have come to rely on hybrid forms of delivery. A research agenda is proposed building on these conceptual frameworks to develop systematic approaches to data collection and analysis to enable clearer and more consistent meanings to emerge.*

Introduction

'it is not just the economy but also the organisations themselves that have become mixed' (Billis, 2010, p. 12)

David Billis neatly summarises a theme that has engaged the interest of a working group of the European Network for Housing Research[1] for a number of years. While the mixed economy of welfare (Powell, 2007) had become an increasingly taken-for-granted outcome of state retreat, privatisation and commissioning of public services from third sector organisations, very little attention seemed to have been paid within housing studies to the implications of this mixing for housing organisations. The European Network of Housing Research (ENHR) working group on 'Social Housing: Institutions, Organisations and

---

*Special issue on social enterprise, hybridity and housing organisations.

Governance' has since 2002 taken a broad interest in change in the social housing sector in Europe, and has explored these issues from a number of perspectives including network and systems theories (Mullins & Rhodes, 2007), marketisation and the introduction of new actors into social housing (Rhodes & Mullins, 2009) and network governance in urban regeneration (Van Bortel *et al.*, 2009). Moreover, the working group became increasingly interested in tensions emerging *within* housing organisations as a result of their blending of market and social goals and mechanisms. Work by Gruis (2008) on the tensions between social and commercial goals in Dutch housing associations and Mullins & Pawson (2010) on hybridity in the English housing sector provided a bridge to the wider literature on social enterprise and hybridity. In this study, we set out to explore the meaning of hybridity and social enterprise in housing organisations by considering how these concepts have been applied in recent studies of different national, sector and organisational contexts.

While increasing rapidly, the literature on social enterprise (see for example Crossan *et al.*, 2003; Dees, 1998; Defourny, 2001; Evers & Laville, 2004; Kerlin, 2006; Lyon & Sepulveda, 2009; Peattie & Morley, 2008; Teasdale, 2011) and on hybridity (see for example Anheier, 2011; Billis, 2010; Brandsen *et al.*, 2005; Evers, 2005; Koppell, 2001; Osborne, 2005; Skelcher, 2004, 2005) has to date provided very few accounts of hybrid models in housing, despite their apparent prevalence in policy and practice. Notable exceptions are Teasdale's (2009, 2010) work on social enterprise models in the homelessness sector and Buckingham's (2009, 2011) work on homelessness support organisations in England. US exceptions include Kopell's (2001) early case studies of Fannie Mae and Freddie Mac in relation to housing finance and Smith's (2010) inclusion of housing in wider work of hybridisation of governance of non-profit organisations. Meanwhile, Blessing has recently provided a theoretically informed account of hybridity and the role of not-for-profit social entrepreneurs as 'magical or monstrous' in the housing market in the Netherlands and Australia (Blessing, 2012, p. 189). The time therefore seemed right for a special issue on social enterprise and hybridity in housing, and we were delighted to receive support from the Housing Studies editorial board for such a proposal.

The result has been an interesting and eventful journey over a two-year period including workshops at two ENHR Conferences in Istanbul in 2010, when some of the papers published here first appeared, and in Toulouse in 2011, when more developed versions of the former papers were subjected to lively discussion and debate both within the workshop and in a special plenary session of the conference. The plenary contributed to our broader aim of drawing attention of the wider body of housing studies scholars to this often overlooked dimension of change in the ways in which housing and community needs are being addressed by new organisational forms and what the concepts of hybridity and social enterprise might mean in these emerging contexts.

This special issue takes this agenda on to the next level by providing an overview of recent international work on social enterprise and hybridity in housing and proposing a future research agenda. We hope that this will provide the stimulus for more scholars to enter this field and to address some of the continued gaps and outstanding questions that endure. We must also remain aware of the critique, that we encountered at times in the referring process, that hybridity is 'a concept that is widely used but seems to play no useful function in theory building or advice to policy-makers' (Skelcher, 2012). Thus, the mere imposition of concepts of hybridity or social enterprise onto empirical investigations may add little to understanding unless they clarify and adequately theorise tensions that would otherwise have been opaque.

The quest for meaning must, of course, start with some attempts at definition. However as another paper in this special issue notes (Czischke *et al.*, pp. 418–437) 'definitions of social enterprise have varied considerably between jurisdictions, and have often been policy dependent rather than subject to more scientific definitions'. Moreover, there has been a reluctance in some places to use the term at all in relation to housing landlord organisations even where they are separate from government, trading for a social purpose and appear to meet policy definitions applied to other fields.

Nevertheless, a minimum definition of social enterprise provided by the UK Department of Trade and Industry of organisations that 'trade for social purpose' (DTI, 2002) appears to have wide currency internationally. Indeed, there is a degree of consensus even across the European and North American divide (Kerlin, 2006) that social enterprise involves the use of non-governmental, market-based approaches to address social issues (Peattie & Morley, 2008). In the USA, the concept remains very broad and dominated by revenue generation in response to government funding cuts (Dees, 1998). Meanwhile in Europe the meaning tends to be more specific, associated with cooperative forms of organisation and is often underpinned by legislation (Defourny, 2009).

Hybridity can be an even more elusive term to pin down, since it often has multiple meanings. The complex nature of hybrid organisations is recognised in Anheier's (2011) view that a necessary condition of hybridity is the presence of relatively persistent multiple stakeholder configurations. Similarly, Billis refers to hybrid organisations as possessing 'significant' characteristics of more than one sector (public, private and third) (Billis, 2010:p. 3). This definition is valuable since it also embraces private sector organisations with a strong social orientation. However, Billis' core interest is in hybrid organisations that have 'roots' and therefore primarily adhere to the 'distinctive principles of just one sector' (*ibid*, p. 3), and in particular to third sector hybrids and their departure from the 'pure form' of the voluntary association.

Other forms of hybridity that are relevant to understanding change in housing organisations include hybrid financial dependencies (mixing state and market funding), hybrid governance structures (reflecting stakeholder mix or separating charitable and commercial activities) and hybrid products and services (combining housing with social and neighbourhood support services). In the social housing context, therefore, Mullins & Pawson (2010) discuss hybridity in English and Dutch housing associations in relation to 'finance, governance, structure and activities' and contrast views of hybrids as 'for profits in disguise or as agents of policy' (2010, p. 197). Meanwhile, Blessing considers 'hybrid' status to imply 'spanning state and market, combining public and private action logics, and subject to multiple sets of institutional conditions' (Blessing, 2012, p.190). Her account of hybridity in the Dutch and Australian rental housing markets conceptualises hybridity alternatively as 'a state of transformation', as providing 'links between cultures', 'hybrid vigour' and 'magical solutions', and as 'transgressing binary divides' between state and market. She concludes that 'social entrepreneurship is not a super-blend, but a balancing act' (Blessing, 2012, p. 205), involving compromises and trade-offs between competing institutional rules and norms.

Key Questions for the Special Issue

The call for papers for this special issue required authors to connect housing studies with the theoretical literature on social enterprise and hybridity and to draw out their relevance for social housing organisations, particularly those that are organisationally separate from government and that are to a degree trading for a social purpose.

Drawing on prior research on gaps in our knowledge on social enterprise and hybridity in the housing sector and our quest for meaning, we identified the following questions which we hoped that the contributions would address:

- How useful are models of social enterprise and hybridity for the analysis of organisational behaviour in the housing sector in different contexts?
- How and to what extent do housing organisations engage with debates about social enterprise and hybridity?
- How do they position themselves vis-à-vis the state, the market and society?
- How do they reconcile conflicting logics of 'common good', financial return and government policy?
- How do these conflicting logics play out in housing policy and implementation in different national and local contexts?
- What are the policy implications of the growth in social enterprise and hybridity?

Contributions

Geographical and Sector Coverage

With the encouragement of the Housing Studies editorial board, we invited contributions from a broader base than the European social housing context that had stimulated our initial interest in social enterprise and hybridity. We are delighted to include in this special issue two contributions from the USA, one focusing on the non-profit housing sector (Bratt, pp. 438–456) and one looking at the emergence of increasingly hybrid forms in the public housing sector (Ngyuen & Rohe, pp. 457–475). We have two papers from the Asia Pacific region, one exploring the nature of hybridity within an expanding and diversifying social housing sector in South Korea (Lee & Ronald, pp. 495–513), the other focused on organisational hybridity in the context of attempts to expand a very small Australian non-profit housing sector (Gilmour & Milligan, pp. 476–494).

Three European contributions include a comparative perspective (Czischke *et al.*, pp. 418–437), which develops an empirically informed conceptual model for a study of social enterprise being applied in a depth study of housing companies in England, Finland and the Netherlands. The two remaining papers focus on England. Teasdale (pp. 514–532) explores the tensions between social and commercial goals in work integration social enterprises in the homelessness field, while Sacranie (pp. 533–552) provides insights into the enactment of hybridity in different parts of a large geographically dispersed housing association in a phase of post-merger integration.

We believe that these papers provide a useful variety of national and organisational contexts in which to explore the meaning of social enterprise, hybridity and housing but recognise that there is scope for considerable expansion of the coverage offered by this collection and for a wider range of interpretations to be considered.

Models and Concepts

It is clear from the papers presented in this special issue that while existing conceptualisations of social enterprise and hybridity are helpful in understanding the dynamics of change in housing organisations in different contexts, there is scope for their refinement to fit specific contexts and to link with broader theoretical perspectives. Peer reviews have also helped us to consider the extent to which such conceptualisation really adds meaning to existing analyses.

Most of the authors have drawn, to some extent, on existing conceptualisations of social enterprise and hybridity to illuminate case studies of housing organisations. These conceptualisations often take on a graphical form; continua, triangles, pyramids and overlapping circles, as illustrations to the papers that follow confirm. In particular, we see social enterprise depicted as a continuum between state and market forms (see for example Crossan, 2009), with the possibility of organisations moving backwards and forwards along this continuum between traditional and more socially entrepreneurial practices (Stull, 2003). We also see repeatedly the formulation of social enterprises as operating in a force field pulled between the three triangular drivers of state, market and 'community' (Brandsen *et al.*, 2005), or in some cases 'society' (Gruis, 2008) or in the case of Buckingham's (2011) pyramid between these forces and a vertical axis representing the third sector itself. Finally, we see reference to a variety of forms of hybridity dependent on the allocation of 'principal ownership' (Billis, 2010) between three overlapping circles of state, market and third sector.

Reference is also made in several papers to Billis' distinction between 'organic hybrids' where organisations moved away from a classical voluntary association form, as a result of key changes such as employment of staff, engagement in trading to generate income and 'enacted hybrids' which are set up from the start as hybrid forms with mixed principal owners (as in the case of stock transfer housing associations in England with their hybrid governance of tenants, local authority persons and 'independents'). The move towards the more entrenched forms of hybridity (defined by Billis, 2010, p. 59) as 'the permanent influence by public and private actors on the governance and operations of an organisation in return for the resources provided by these actors' is evident for example in US public housing organisations where the entire operation has become dependent on resources from a combination of sectors and could no longer functions as a 'pure' single sector organisation (Ngyuen and Rohe). *resource dependency theory.*

Prompted by at least one reviewer and by some of the more sceptical literature (Skelcher, 2012), we asked ourselves the question how much the use of hybridity and social enterprise concepts added meaning to our understanding of the case studies presented in this special issue. While the continua, triangles and overlapping circles discussed above have a value in depicting different mixes of state, market and societal influences on organisations, we believe that these conceptualisations have been most useful in highlighting the directions of change of organisations and sectors and pinpointing the underlying drivers of change rather than simply to indicate organisational positioning. The conflicting meanings associated with state, market and community drivers have informed analyses of organisational changes in several of the papers. Czischke *et al.* (pp. 418–437), in particular argue that it is important to understand how competing hybrid principles are applied in organisational strategies and decisions. They develop a framework incorporating descriptor, motivator and behaviour variables to track the ways in which hybridity shapes organisational behaviour in specific decision contexts. The shifting boundaries between state, market and communities around public housing are apparent in several of the cases, particularly in the account of hybridisation of public housing in the USA (Ngyuen & Rohe, pp. 457–475). The contests between state, market and community influences on a single organisation's community investment strategies are captured by Sacranie's (pp. 533–552) analysis of a large merged English housing association.

The authors also offer a variety of additional conceptualisations to enhance the understanding of their case studies. In most cases, these concepts are compatible with, rather than competing with, social enterprise and hybridity frameworks. For example, Bratt's (pp. 438–456) development of her earlier 'quadruple bottom line' construct informs her analysis of hybridisation arising from the increasing complexity of tasks taken on by US non-profits. Gilmour & Milligan (pp. 476–494) propose the use of institutional theory to understand the low levels of isomorphism among the different types of non-profit housing providers that have emerged in Australia from the diverse financing models and strategic orientation of hybrid organisations. More explicit use of institutional theory is made by Sacranie (pp. 533–552) who draws on literature on institutional logics as well as organisational cultures to track the enactment of hybridity in a large English housing association. Teasdale (pp. 514–532) invokes the concept of resource transfers to explore how trade-offs between commercial and social goals are transformed from a zero sum game in homeless social enterprises. Lee & Ronald (pp. 495–513) place their analysis of hybridisation within an understanding of developmental welfare regimes in East Asia and their specific path dependency. This helps to explain the continued role of state steering and regulation by the Korea Land and Housing Corporation and the emergence of hybridity as part of an enlargement of public housing to accommodate more non-market forms and a wider range of housing-related services (housing welfare centres, Chonsie loans, etc.) rather than as the expansion of an independent privately driven sector. While not conflicting with theorisations of hybridity and social enterprise, these alternative theoretical lenses do help to explain dimensions of the case studies and to draw out meaning in deeper ways and thereby to overcome the criticism that hybridity alone can be seen as a rather abstract label that fails to illuminate underlying change processes.

By focusing on dynamic processes of hybridisation rather than static descriptions of hybridity, by setting these processes in a broader social and political context and by identifying underlying change mechanisms such as competing organisational logics, trade-offs between social and commercial goals and resource transfers, the papers in this special issue have risen to the challenge of more critical theorisation that adds meaning to the case studies they explore.

Policy Drivers

Most of the case study contexts presented here demonstrate the existence of policy drivers that are leading housing organisations to adopt more hybrid business models. In the USA and recent Australian policy frameworks, we find explicit financial drivers for hybridity in tax credit systems. Gilmour and Milligan argue that the similarities between the Australian National Rental Affordability Scheme 2008 and the US tax credits system are leading to similar organisational practices in the two countries, for example in the tendency to outsource tenancy management to not-for-profit or private sector companies. In the US context, Smith has discussed the ways in which trading activities adopted by non-profits in response to policy drivers can in turn lead to hybrid governance structures (e.g. to accommodate charitable funding, tax status and joint ventures with the private sector). In England, the longstanding policy of treating borrowing by non-profit housing organisations in a different way to state housing led to the increasing adoption of hybrid governance models through stock transfer shifting the balance on the English housing sector towards enacted hybridity (Mullins & Pawson, 2010).

As Teasdale (pp. 514–532) outlines, the English homelessness sector has been subject to a stream of policy initiatives that have sought to promote social enterprise models as a way to help to reintegrate homeless people into the labour market and thereby combat social exclusion. This policy pressure to adopt social enterprise models has been pervasive across the third sector (Harris, 2010) and sometimes been seen as coercive (Carmel & Harlock, 2008). While the policy motivation in the homelessness sector is rather different from financially driven measures to replace public funding with private funding, the consequences have been similar in increasing hybridity in some homelessness organisations between trading arms that are expected to generate a surplus that can be reinvested in the business or used for other mission-related purposes.

The general policy driver advancing hybrid structures in North-Western Europe, the US and Australia have been associated with attempts to substitute private for state expenditure in the context of residualisation of social housing and decline of welfare states. However, Lee and Ronald draw our attention to a very different context for hybridisation in Korea where a developmental state has led to a different role for the social housing sector with diversification and new forms of provision to serve the needs of various types of household as part of an expansion of public housing. They draw attention to similar developments in China identified by Wang & Murie (2011) where, despite intensive housing marketisation, non-market housing options have also been advanced by the state resulting in another 'hybrid system' that reflects traditional legacies and indigenous processes of urbanisation and demographic change.

The consequences for policy of these hybridisation processes are discussed in the concluding section of this editorial.

Conflicting Logics and Cultures

One of the most compelling reasons for considering change in housing organisations through the lens of hybridity and social enterprise is to capture the underlying tensions associated with competing institutional logics (Mullins, 2006; Scott, 2001; Thornton, 2004). These tensions are brought out most strongly in two of the papers, by Teasdale in relation to involving homeless people directly in commercial activities of work integration social enterprises and by Sacranie in relation to a locally based subsidiary within a large English housing group. The playing out of the four competing dimensions (financial viability, social and economic needs of residents, neighbourhood viability and environmental sustainability) of the quadruple bottom line is nicely explored by Bratt drawing on a variety of case study material from the US non-profit housing sector.

Teasdale shows that the supposedly complementary aims of trading and combating social exclusion of homeless service users by providing employment opportunities are often in conflict in practice and there is a trade-off between social and commercial considerations. Six case study organisations were found to manage these trade-offs in different ways, usually by bringing in additional resources from other sectors of the economy to remain competitive. None were able to bring homeless people into employment and provide the necessary support using commercial revenue alone. Instead they harnessed their hybrid identities to secure additional resources including charitable donations, volunteer labour, ethical consumers willing to pay a price premium for goods and services delivered by homeless people, state benefits and payments for employment-related services. Moreover, the tensions between involving homeless people (many of whom may be less economically productive) in commercial activities and providing them with the necessary support were often managed by separating the trading and support arms of these organisations into separate and loosely coupled activities.

Sacranie's analysis depicts a process of contestation within a large hybrid organisation between competing institutional logics. The first logic is a 'customer' commercial business logic tied to the corporate and centralised culture of the dominant partner; meanwhile a competing 'community' logic is found in some of the business units including a former black and minority ethnic (BME) housing association with a community outreach programme. The paper tracks the displacement of the latter logic by the former within the BME housing association and the shifts in resource dependencies and governance and staff changes that underpinned this transformation. This shift is most apparent in decision making on priorities for community investment for which the focus has shifted from a locally responsive community partnership approach to a corporate strategic approach that one might expect to find in large-scale private sector business.

Moving up to the sector level, Bratt's paper depicts a change process in which over time US non-profit housing organisations with specific and diverse origins in responding to places, people or projects began to broaden their missions to embrace multiple goals. Increasing complexity then accentuated competition between the different components of the quadruple bottom line mentioned earlier. This competition was further accentuated by federal funding reductions leading to increasingly hybrid responses, such as structures designed to facilitate the exploitation of tax breaks under the Low Income Housing Tax Credit Programme which 'pushed nonprofits to ever greater degrees of hybridity' (p. 449). Bratt points out the irony of a hybridity driven by the end of deep public subsidies, in that non-profits were forced into ever closer links with 'the very actors and institutions that have neglected the nonprofits' target constituents and communities' (*ibid*, p. 450). Moreover, Bratt sees the process of hybridisation as thwarting their ability to meet each dimension of the quadruple bottom line, for example Community Development Corporations have become disconnected from their constituents and lost their advocacy roles as they have become enmeshed in financial dependence on the private sector.

Directions for Future Research

While drawing on a limited range of contexts and case studies, we believe that this special issue provides a useful staging post for further work on social enterprise and hybrid models in housing organisations. The coherence of the special issue has benefited from the call to include theoretical as well as empirical contributions and to address some common questions. However, it is recognised that the authors were generally drawing on research and data collected with slightly different research questions in mind. Future comparative research should benefit by building on the theoretical and conceptual frameworks suggested here and by developing a systematic approach to data collection and analysis. In this way we will be clearer that the meanings applied to concepts and the resulting meaning of and interpretations of change will be more consistent and coherent.

There are a range of ways in which such a systematic approach could be developed to housing studies research on social enterprise and hybridity.

First, the editors are involved in a series of national Delphi panel studies in which perspectives on sector change are being constructed in partnership with leaders of third sector housing organisations engaged in balancing market, state and community influences on decisions on a day-to-day basis.[2] These studies are using a core of common questions within survey instruments tailored to reflect the specific policy, institutional and frameworks in operation in each country. There is scope for this approach to be adopted in a wider range of countries to build our understanding of how hybrid drivers are being enacted within organisations in different contexts internationally. In this special issue, the differences of context in South Korea, Australia, the USA and Northern Europe have been clear, indeed the case is well made by Lee and Ronald that Eurocentric conceptual models of social enterprise need to be tested and refined in other contexts such as South Korea and China, where social housing has been expanding rather than contracting. However, the value of considering a set of common questions to understand the different impacts of hybridity at the organisational level in different contexts is also evident.

Second, there is scope for meta-reviews of existing research of social enterprise and hybrid models in housing drawing on a number of current PhDs and research projects known to be in progress in this field. By conducting such meta-reviews with closer reference to emerging theory and empirical research within in the wider field of study of social enterprise and hybridity, there is scope for housing studies to make a stronger contribution. In this special issue, the relevance of research on work integration social enterprises in the homelessness sector (Teasdale, pp. 514–532) in understanding tensions arising from involvement of beneficiaries in income generation is clear for housing organisations also involved in supporting trading activities for residents such as grounds maintenance and handyman services. The relevance to analysis of programmes such as HOPE VI and Moving to Work in shifting the organisation of public housing providers into more hybrid forms of organisation and service delivery (Ngyuen & Rohe, pp. 457–475) is apparent.

Third, there is a need for more longitudinal studies such as that undertaken by Sacranie in understanding the enactment of hybridisation processes in housing organisations. Her depiction of the hybridisation process in a large housing association following a significant merger as a contest between competing logics has clear potential for replication and extension, perhaps through ethnographic or participant observation research grounded in individual organisations. Furthermore, the conceptual framework developed by Czischke *et al.* (pp. 418–437) following Crossan (2007) and Stull (2003) provides a template for data collection for organisational case studies to inform further comparative research both nationally and supra-nationally. The latter has been applied by one of the authors in an on-going PhD. This may shed light on the substantive commonalities and differences (motivator and behaviour variables) between social housing providers in North-West Europe beyond legal and/or organisational features (descriptor variables), thereby informing policy making at European level. This would mirror similar comparative research already been conducted at European level on other types of social services[3] childcare, work integration, etc. Suggested methods of exploring the longitudinal process of change include the study of 'critical incidents', strategic decision diaries and stakeholder analyses.

Fourth, critiques of the underlying assumptions of hybrid models and their ability to deliver both commercial and social returns need to be developed in varying economic and social contexts. Bratt's quadruple bottom line clarifies the more complex trade-offs facing

US non-profit housing organisations. Teasdale's paper is particularly useful in confronting the implied assumption that both social and commercial returns can be maximised rather than traded off against each other. Strategies to make such trade-offs in practice by work integration social enterprises in the homelessness field by levering in resources from other sectors of the economy (from government, philanthropists, consumers and other partners) highlight the need to consider interorganisational relations when considering how blended value can be delivered in practice. This begins to address the concern raised by several reviewers of the papers of the extent to which, and under what conditions, social enterprise models can substitute for public subsidy for social and affordable housing rather than simply changing the way in which such subsidy is channelled. In the challenging global conditions that have prevailed since 2008, the continued ability of asset-based models to cross-subsidise social returns from asset value deserves much closer scrutiny.

What are the Policy Implications of the Growth in Social Enterprise and Hybridity?

Earlier in this editorial we explored some of the **policy drivers** for hybridisation and the adoption of social enterprise models. The bigger question, that some papers have begun to explore and where more work is urgently needed, concerns the **policy consequences** of housing systems that become increasingly reliant on hybrid forms of delivery. These are the questions which will give deeper meaning to the concepts by defining some of the limits to their operation in a changing policy context.

In the European and Australian papers, there was an indication of the policy tensions surrounding cross-subsidy underpinnings of those hybrid models that offer the prospect of substituting commercial income for all or part of state contributions to fund social housing. These policy consequences were highlighted by changes in the market and state regulatory contexts that became increasingly apparent during the period in which we were working on the special issue.

The credit crisis and recession of the late 2000s posed severe limitations on cross-subsidy, for example in the property development process and in the ability of housing social enterprises to harness their assets and to build housing for sale to cross-subsidise housing for low-income groups and wider community investment activities to make such housing sustainable.

Political and regulatory barriers have also increased in prominence. We have seen some of the limits imposed on this model in the Netherlands where EU intervention in support of competition policy has forced the separation of commercial and social arms of housing associations and undermined cross-subsidy approaches (Gruis & Primeus, 2008). We are seeing similar tensions in England where charitable status is threatened when commercial activities exceed a *de minimis* threshold even where surpluses from such activities are clearly reinvested in social housing (McDermont, 2010).

The Korean experience (Lee & Ronald, pp. 495–513) seems to suggest that it is possible for a degree of decentralisation to hybrid agencies to coexist with high levels of state steering and regulation. Indeed as the nascent, non-profit housing sector in Australia highlights (Gilmour & Milligan, pp. 476–494), expansion and scaling up of third sector housing often leads to a ramping up of regulation (in the Australian case from state to federal levels to accommodate national growth providers and potential institutional funders). It will be interesting to see whether recent state-led initiatives to exploit commercial strengths of English housing associations to cross-subsidise new building through the move to near market rents, asset sales and investment of accumulated surpluses under the 'new investment framework' will be affected by similar constraints. However, the coincidence of this new investment framework with a drastic cutback in regulation [abolition of the Tenant Services Agency and retreat to a financial back-stop form of regulation within the Homes and Communities (funding) Agency] may pose new challenges. It is instructive in this respect to end on a cautionary note from the wider experience of hybrid housing finance institutions in the US context.

Kopell's (2001) perceptive earlier analysis of Fannie Mae and Freddie Mac had highlighted the risks associated with such hybrid government sponsored agencies (GSEs) that appeared to be beyond the control of government:

> The infrastructure to 'control' them may exist, but GSEs have the resources, ability, and position to control their own controllers... Few would have anticipated that Fannie Mae and Freddie Mac would grow into political heavyweights.... As proposals for additional GSEs are considered, it would be wise to consider the consequences of creating independent political actors with potential influence greater than most institutions, public or private (*ibid*, p. 478).

The subsequent rolling out of the credit crisis and the implication of Fannie Mae and Freddie Mac in the uncontrolled expansion of mortgage credit that initiated it has added to the more modest aspirations for social enterprise models in the austere conditions in the second decade of the twenty-first century.

Perhaps it is by considering the operational tensions and limitations that have emerged in recent troubled times that the meaning of social enterprise and hybridity in housing organisations will be more clearly understood.

Acknowledgements

Thanks to all of the contributors to the Housing Studies special issue and to the workshops at ENHR conferences in 2010 and 2011 who helped to develop and test the ideas on which this paper draws. Particular thanks to Hal Pawson and the Housing Studies editorial board for unstinting support to the special issue editors throughout the peer review process.

Notes

[1] The ENHR working group on '*Social Housing: Institutions, Organisations and Governance*' is currently coordinated by the three editors of this special issue together with Tony Gilmour (Swinburne University, Australia). It has held workshops at nine conferences of the European Network for Housing Research commencing in Vienna in 2002 and has maintained a focus on the organisational and interorganisational level in documenting and analysing change of relevance to housing systems and housing policy research.

[2] See, for example Mullins (2006), Nieboer & Gruis (2011) and Milligan et al. (2012, forthcoming).

[3] For examples of these research projects, see http://www.emes.net

References

Anheier, H. (2011) *Governance and Leadership in Hybrid Organisations. Comparative and Interdisciplinary Perspectives*, pp. 1–7 Background paper (Heidelberg: Centre for Social Investment. University of Heidelberg).

Billis, D. (Ed.) (2010) *Hybrid Organizations and the Third Sector. Challenges for Practice, Theory and Policy* (Basingstoke: Palgrave Macmillan).

Blessing, A. (2012) Magical or monstrous? Hybridity in social housing governance, *Housing Studies*, 27(2), pp. 189–207.

Brandsen, T., van de Donk, W. & Putters, K. (2005) Griffins or chameleons? Hybridity as a permanent and inevitable characteristic of the third sector, *International Journal of Public Administration*, 28, pp. 749–776.

Buckingham, H. (2009) Competition and contracts in the voluntary sector: Exploring the implications for homelessness service providers in Southampton, *Policy and Politics*, 37(2), pp. 235–254.

Buckingham, H. (2011) Hybridity, diversity and the division of labour in the third sector: What can we learn from homelessness organisations in the UK? *Voluntary Sector Review*, 2(2), pp. 157–175.

Carmel, E. & Horlock, J. (2008) Instituting the 'third sector' as a governable terrain: Partnership, performance and procurement in the UK, *Policy and Politics*, 36(2), pp. 155–171.

Crossan, D., Bell, J. & Ibbotson, P. (2003) *Towards a Classification Framework for Social Enterprises*. Paper presented at the ARNOVA Conference 2003.

Crossan, D. (2007) Towards a classification framework for not for profit organisations, PhD Dissertation Thesis, School of International Business, University of Ulster, Magee Campus.

Crossan, D. & Til, J.V. (2009) Towards a classification framework for not-for profit organisations - the importance of measurement indicators. EMES Conferences Selected Papers, pp. 1–25.

Dees, J. G. (1998) Enterprising nonprofits, *Harvard Business Review*, 76(1), pp. 55–675.

Defourny, J. (2001) Introduction: From third sector to social enterprise, in: C. Borzaga & J. Defourny (Eds) *The Emergence of Social Enterprise*, pp. 1–28 (London/New York: Routledge).

Defourny, J. (2009) Concepts and realities of social enterprise: A European perspective, in: *Collegium*, No. 38, Spring2009. Special edition.

DTI, (2002) *Social Enterprise Strategy for Success*, Department of Trade and Industry (DTI), London.

Evers, A. (2005) Mixed welfare systems and hybrid organizations: Changes in the governance and provision of social services, *International Journal of Public Administration*, 28, pp. 737–748.

Evers, A. & Laville, J.-L. (2004) Defining the third sector in Europe, in: A. Evers & J.-L. Laville (Eds) *The Third Sector in Europe*, pp. 11–30 (Cheltenham: Edward Elgar).

Gruis, V. (2008) Organisational archetypes for Dutch housing associations, *Environment and Planning C: Government and Policy*, 26(6), pp. 1077–1092.

Gruis, V. & Primeus, H. (2008) European competition policy and national housing policies: International implications of the Dutch case, *Housing Studies*, 23(3), pp. 485–505.

Harris, M. (2010) Third sector organisations in a contradictory policy environment, in: D. Billis (Ed.) *Hybrid Organizations and the Third Sector. Challenges for Practice, Theory and Policy*, pp. 25–45 (Basingstoke: Palgrave Macmillan), Chapter 2.

Kerlin, J. A. (2006) Social enterprise in the United States and Europe: Understanding and learning from the differences, *Voluntas*, 17, pp. 247–263.

Koppell, J. G. S. (2001) Hybrid organizations and the alignment of interests: The case of Fannie Mae and Freddie Mac, *Public Administration Review*, 61(4), pp. 468–482.

Lyon, F. & Sepulveda, L. (2009) Mapping social enterprise: Past approaches, challenges and future directions, *Social Enterprise Journal*, 5(1), pp. 83–94.

McDermont, M. (2010) *Governing Independence and Expertise: The Business of Housing Associations* (Oxford: Hart Publishing).

Milligan, V., Nieboer, N., Hilse, K. & Mullins, D. (2012) The old and the new: Comparing strategic positioning of third sector housing organisations in the Netherlands and Australia, ENHR Conference Lillehammer, Norway.

Mullins, D. (2006) Exploring change in the housing association sector in England using the Delphi method, *Housing Studies*, 21(2), pp. 227–251.

Mullins, D. & Pawson, H. (2010) Housing associations: Agents of policy or profits in disguise? in: D. Billis (Ed.) *Hybrid Organizations and the Third Sector. Challenges for Practice, Theory and Policy*, pp. 197–218 (Basingstoke: Palgrave Macmillan), Chapter 10.

Mullins, D. & Rhodes, M. L. (2007) Editorial, special issue on network theory and housing systems, *Housing Theory and Society*, 24(1), pp. 1–13.

Nieboer, N. & Gruis, V. (2011) Shifting back in the Dutch social housing sector, ENHR Conference 2011, July 5–8, Toulouse.

Osborne, S. P. (2005) Voluntary action in a changing Europe: Critical perspectives, *International Journal of Public Administration*, 28, pp. 733–735.

Peattie, K. & Morley, A. (2008) *Social Enterprises: Diversity and Dynamics, Contexts and Contributions* (London: Social Enterprise Coalition).

Powell, M. A. (2007) *Understanding the Mixed Economy of Welfare* (Bristol: Policy Press).

Rhodes, M. L. & Mullins, D. (2009) Editorial special issue on market concepts, coordination mechanisms and new actors in social housing, *European Journal of Housing Policy*, 9(2), pp. 107–119.

Scott, W. R. (2001) *Institutions and Organisations*, 2nd ed. (London: Sage).

Skelcher, C. (2004) Hybrids: Implications of new corporate forms for public service performance, Paper presented at British Academy of Management Conference, Edinburgh.

Skelcher, C. (2005) Public–private partnerships and Hybridity, in: Ewan Ferlie, Lawrence E. Jr, Lynn & Christopher Pollitt (Eds) *Oxford Handbook of Public Management*, pp. 347–370 (Oxford: Oxford University Press).

Skelcher, C. (2012) *What do we mean when we talk about 'hybrids' and 'hybridity' in public management and governance?* Working Paper. pp. 1–27. University of Birmingham, Institute of Local Government Studies. Unpublished. Available at http://epapers.bham.ac.uk/1601/1/Hybrids_working_paper_march_2011.pdf

Smith, Stephen. Rathgeb (2010) Hybridization and Nonprofit Organizations: The Governance Challenge, *Policy and Society* 29(3), pp. 219–229.

Stull, M. G. (2003) Balancing the dynamic tension between traditional nonprofit management and social entrepreneurship, Paper presented at the ARNOVA 2003 Conference.

Teasdale, S. (2009) *Innovation in the homelessness field: how does social enterprise respond to the needs of homeless people?*, TSRC Working Paper 5 (Birmingham: University of Birmingham).

Teasdale, S. (2010) Models of social enterprise in the homelessness field, *Social Enterprise Journal*, 6(1), pp. 23–34.

Teasdale, S. (2011) What's in a name? Making sense of social enterprise discourses, *Public Policy and Administration*, 27, pp. 99–119.

Thornton, P. H. (2004) *Markets from Culture. Institutional Logics and Organisational Decisions in Higher Education Publishing* (Stanford, US: Stanford University Press).

Van Bortel, G., Mullins, D. & Rhodes, M. L. (2009) Editorial, special issue on exploring network governance in urban regeneration, community involvement and integration, *Journal of Housing and the Built Environment*, 24(2), pp. 93–101.

Wang, Y. P. & Murie, A. (2011) The new affordable and social housing provision system in China: Implications for comparative housing studies, *International Journal of Housing Policy*, 10(3), pp. 237–254.

Conceptualising Social Enterprise in Housing Organisations

DARINKA CZISCHKE*, VINCENT GRUIS* & DAVID MULLINS**
*Faculty of Architecture, Department of Real Estate and Housing, Delft University of Technology, Delft, The Netherlands, **School of Social Policy, University of Birmingham, Birmingham, UK

ABSTRACT   *Recent changes in the provision, funding and management of social housing in Europe have led to the emergence of new types of providers. While some of them can be portrayed with traditional 'state', 'market' or 'civil society' labels, many correspond to hybrid organisational forms, encompassing characteristics of all three in varying combinations. Nonetheless, evidence suggests that there is a 'common thread' linking these organisations together, namely their core missions and values, which can be classified using the term 'social enterprise'. Despite the growing body of literature on social enterprise, this concept has been poorly defined and applied to the housing sector. This paper aims to address this gap through a critical literature review encompassing Europe and the United States. Existing models of social enterprise are reviewed and a classification system for social enterprise is developed to reflect the specific features of the social housing association sector and as framework for future research.*

Introduction

Over the past decades, a number of changes in the provision, funding and management of social housing in Europe have been commonly described as a process of 'privatisation' (e.g. Czischke & Gruis, 2007; Priemus *et al.*, 1999). On closer examination this process has taken a number of forms and has involved alongside the profit distributing private sector a number of other sets of organisations, often new landlords balancing social and commercial objectives (Pawson & Mullins, 2010). The key set of organisations considered in this paper are social purpose organisations operating on a non-profit distribution basis, constituting a wide range of 'third sector' actors providing social and affordable housing across Europe. Although different national legal and institutional frameworks shape the scope and field of their activities, an explorative review by Heino *et al.* (2007) suggested that there is a 'common thread' linking these organisations together, namely their core missions and values, which can be classified under the more general heading of 'social enterprise' (e.g. Boelhouwer, 1999; Gruis, 2008) or related concepts such as 'social economy enterprises (Mertens, 1999), 'businesses for social purpose' (Mullins & Sacranie, 2009) and hybrid organisations (Billis, 2010; Mullins & Pawson, 2010). Indeed, while some of these organisations can be portrayed with traditional 'state', 'market' or 'civil society' labels, many correspond to hybrid organisational forms, encompassing characteristics of the three in varying combinations (Evers, 2005; Evers & Laville, 2004). However, despite the growing body of literature in the field of social entrepreneurship, with a few exceptions such as in the homelessness field (Buckingham, 2010; Teasdale, 2009a, 2009b), and comparative work on the state, market and community models for social housing in the United States, Australia and the UK (Gilmour, 2009), this concept has been poorly developed in relation to housing studies. Furthermore, we maintain that the concept of social enterprise has been theorised differently in different countries and sectoral contexts (Brouard, 2007; Crossan & Til, 2009; Defourny, 2009), thereby leading to important distinctions, which must be considered in the application of this concept to the field of (social) housing. As Crossan & Til (2009) point out better knowledge of this type of organisation is needed to improve policy formulation in this field, a sentiment followed by Buckingham in relation to homeless support commissioning (2010). This paper addresses this gap by developing a classification framework for social enterprise to apply to social housing organisations.

Our conceptualisation of social enterprise in housing draws on a review of general classification models for social enterprise. Existing literature on 'social enterprise' and related concepts are reviewed to establish core characteristics that are applicable to housing studies. Differences in conceptualisation across Europe are explored. Using the more general theoretical classifications of social enterprise as well as some preliminary work from the housing sector, a framework is presented to classify various social enterprise approaches within social housing organisations. Thereby, this paper aims to shed light on a variety of semantic distinctions and critically assess the applicability of the existing literature and conceptualisation of social enterprise to facilitate the study of social housing organisations.

This paper is structured as follows: the second section explores the concept of 'social enterprise' and discusses several general classification models. Section three presents a proposed classification framework for social enterprise in housing organisations, emerging from the literature review. Section four sums up the conclusions, including a critical discussion of the applicability of the adapted classification model to social housing organisations and directions for further research.

The Concept of Social Enterprise

The term 'social enterprise' is part of a large family of inter-related concepts. In fact, it is often used in connection or even interchangeably with other terms such as 'social economy', 'business for social purpose', 'not-for-profit organisations', 'third sector organisations', 'voluntary organisations', etc. Similarly in the case of other terms, such as 'social capital' or 'corporate social responsibility', definitions are manifold and semantic boundaries tend to be somewhat fuzzy. Nonetheless, there seems to be a consensus among different authors studying this phenomenon in that 'social enterprise' and other related concepts have emerged within the umbrella of the notion of 'third sector', a '(...) blanket definition [encompassing] all the small-scale production units set up by individuals or community groups with a view to trying out novel collective working practices and to filling a hole as regards meeting a genuine need' (Mertens, 1999, p. 502). The term 'third sector' is deemed to have gained popularity in France in the late 1970s. Delors & Baudin (1979) is credited as the first to try and quantify the phenomenon, describing it as a 'variation on the theme of the services sector' (Bidet, 1997, p. 62), coexisting alongside the market economy and the state sector. Indeed, as Mertens points out, today '(. . .) most researchers accept the definition of "a collection of organisations which are neither capitalist nor run by the state"' (1999, p. 502). Evers (2005) refers to this diversity of providers as the 'welfare mix', and depicts third sector welfare providers as operating within a triangle bounded by state welfare, market welfare and informal welfare systems and sharing some characteristics of each. A similar framework developed by Brandsen *et al.* (2005) is discussed later (see Figure 3).

Moreover, as mentioned earlier, national differences in the way the concept is used add to this semantic complexity. In his account of the origins and definition of the term 'social enterprise', Defourny explains that the concept has been gaining importance for the past decades on both sides of the Atlantic albeit with very different meanings:

> (. . .) the social enterprise concept remains very broad in the U.S. (. . .) A dominant trend consists in using the term to designate market-oriented economic activities serving a social goal. Social entrepreneurship is then viewed as an innovative response to the funding problems of non-profit organisations, which are finding it increasingly difficult to solicit private donations and government and foundation grants. (. . .) the Anglo-Saxon concept (especially in the U.S.) remains characterised by a profusion of approaches and definitions, mainly around major 'business schools' and management sciences in general. (Defourny, 2009, p. 74)

For example, Hasenfeld & Gidron (2005) refer to civil society, social movements and non-profit organisations for understanding multi-purpose hybrid voluntary organisations. In a paper comparing social enterprise in the United States and Europe, Kerlin points out that the concept in the former is 'generally much broader and more focused on enterprise for the sake of revenue generation than definitions elsewhere' (2006, p. 248). Furthermore, she refers to a definitional divide in the United States. between academics and practitioners:

> In U.S. academic circles, social enterprise is understood to include those organisations that fall along a continuum from profit-oriented businesses engaged in socially beneficial activities (corporate philanthropies or corporate social responsibility) to dual-purpose businesses that mediate profit goals with social objectives (hybrids) to non profit organisations engaged in mission-supporting commercial activity (social purpose organisations). For social purpose organisations, mission-supporting commercial activity may include only revenue generation that supports other programming in the non profit or activities that simultaneously generate revenue and provide programming that meets mission goals such as sheltered workshops for the disabled (Young, 2001, 2003). (2006, p. 248)

Meanwhile, in Europe, as Defourny (2009) points out, despite the conceptual diversity around the term 'social enterprise', this notion seems to have appeared earlier than in the United States and with more specific meanings often underpinned by legislation. For example, Defourny sees the adoption in 1991 of a law giving specific status to 'social co-operatives' by the Italian Parliament as an important milestone followed by significant growth of this type of organisations. Interestingly, these co-operatives had originally emerged as a response to needs that had either not been met or inadequately been met by public services, which coincides with the definition of 'third sector' seen above. Apart from the law, authors on both sides of the Atlantic have identified a number of macro-structural factors such as persistent unemployment, decrease of public funding and increasing (and unmet) social needs that have as a response given rise to the proliferation of third sector actors (Brouard, 2007; Defourny, 2009). Another key difference in usage is between social enterprise as a set of activities developed in response to the types of needs outlined above, and social enterprise as an organisational form. We will return to this distinction in the following section.

The Dutch Network of Social Enterprises (NTMO, 2003) defines social enterprises as organisations that have been designed as private enterprises, operating in a market situation, that at the same time use their means to fulfil a societal objective that is interwoven with (or parallel to) the common interest, that produces goods and services and that uses its profit entirely for the realisation of its societal objective (contrast this with the UK definition below—principally rather than entirely). This widely accepted definition of social enterprise in the Netherlands is similar to the US definition. In addition, as de Boer's definition (1999, p. 20) points out, social enterprise in the Netherlands is not only used to refer to formal institutional characteristics, but also to identify a specific organisational approach:

> Social enterprise is mobilising the force of entrepreneurship for the public cause. Against this background, a social enterprise can be described as a private, not-for-profit organisation that attempts to use public and private means to realise public objectives and, in doing so, uses principles from commercial enterprises such as innovation, market orientation and taking risks.

This concept of social enterprise has been widely and explicitly adopted in the Dutch housing association sector as well. Since, as a result of neo-liberal policies, Dutch housing associations have gained considerable administrative and financial independence in the 1990s, social enterprise has been adopted as a concept to refer to their relatively independent position, while retaining the social purpose of the organisations. Furthermore, the concept has been used to refer to the adoption of entrepreneurial behaviour and strategies within the housing associations to deal with social housing and related objectives in a more effective and efficient way. More specifically, social enterprise has also been related to specific shapes of corporate governance within housing associations, including explicit strategies for stakeholder involvement in decision-making and accountability towards stakeholders (e.g. de Kam, 2003; Gruis, 2010; SER, 2005; van Dijk *et al.*, 2002; Vulperhorst, 1999; Zandstra & Rohde, 2002).

In England there have been a variety of definitions of social enterprise reflecting political and policy drivers; and particular issues have arisen in defining 'social aims' and 'trading income'; the twin planks of the concept (Lyon & Sepulveda, 2009). A very open definition was adopted by the government in 2002, responding to the London-based Social Economy Coalition (which grew out of the co-operative sector) as part of its strategy 'Social Enterprise: Strategy for Success' 'A social enterprise is a business with primarily social objectives whose surpluses are principally reinvested for that purpose in the business or the community, rather than being driven by the need to maximise profits for shareholders and owners'. (DTI, 2002). This definition does not translate into a single legal or regulatory form, although as part of the governmental promotion strategy a new legal form, Community Interest Companies, was created and has been adopted by a small sub-group of organisations. The wide diversity of organisational types captured by the definition includes 'development trusts, community enterprises, housing associations, football supporter's trusts, social firms, leisure trusts and co-operatives' (Office for the Third Sector, 2006). Moreover, Lyon and Sepulveda have noted that the broad definitions mean that 'many organisations that do not define themselves as social enterprises are defined as such, but would agree that they are involved in social enterprise activity' (2009b, p. 3). This is certainly the case in relation to most English housing associations which, in contrast to Dutch associations tend to talk about social enterprise as something they *support* (e.g. tenant-led businesses) rather than something they *are* themselves as the following example from the National Housing Federation Conference, 2010 brochure illustrates 'be inspired and find out the role your association can play, either in establishing a social enterprise or fostering partnerships to achieve more for your communities' (NHF, 2010, p. 7).

Table 1 shows some key milestones in the emergence and progressive adoption of the concept of 'social enterprise' since the mid seventies both in the United States. and Europe. It is worth noting, however, that long-standing traditions of mutual organisations and co-operatives have existed for over a century in Europe (Defourny, 2009). Furthermore, Defourny distinguishes two conceptual approaches aiming to encompass the

Table 1. Milestones in the origin and development of the concept of 'social enterprise' in Western Europe.

Year	Europe
1979	France: Jacques Delors coins the term 'third sector'.
1990	Italy: Journal *Impresa sociale* starts studying new entrepreneurial dynamics with a social purpose.
1991	Italy: Italian Parliament adopts law conferring specific status to 'social co-operatives'.
1996	European researchers form a network to study the emergence of social enterprises in Europe (EMES).
2001	United Kingdom: Social Enterprise Coalition influences Blair Government which responds by creation of a 'Social Enterprise Unit' within Department of Trade and Industry to improve knowledge on social enterprises and promote them throughout the country.
2004	United Kingdom: *Companies Act* 2004 establishes new legal form for social enterprise—Community Interest Companies Creation of Office for Third Sector in Cabinet Office gives prominent role to expanding social enterprise in public service provision. Labour Government 'Created a new third sector and placed it at the centre of a new third way for policy development' (Alcock, 2010, p. 14).

Source: Build on Defourny (2009); Mertens (1999); Alcock (2010).

whole third sector, namely the 'non-profit sector' approach (mainly developed in the English-speaking world), and the 'social economy' approach, of French origin (Mertens, 1999). The latter brings together co-operatives, mutual societies, associations and foundations. Both approaches have gradually spread internationally, along with statistical work seeking to quantify the economic importance of the sector.

Within this wider conceptual framework, the notion of 'social enterprise' appears to signal a step forward in relation to the 'traditional' co-operatives or social economy organisations, in that it implies a 'new way of doing things'. This idea connects with concepts such as (social) innovation and styles of management seeking greater efficiency in service delivery, similar to those that can be found in the profit-oriented sector (Mertens, 1999). Indeed, as Defourny points out (2009), the recent introduction of new legal frameworks for these types of initiatives in different European countries seems to confirm the emergence of this new type of entrepreneurship vis-à-vis more traditional approaches. A key question arising from this work is the extent to which social enterprise is seen as substituting for the state or complementing the state; especially in relation to public expenditure and alternative sources of income. While the notion of trading appears to imply reduced state dependence, this may not be the case if the trading is principally for governmental contracts.

Classification Models for Social Enterprise

As explained in the previous section, the definition of social enterprise varies significantly across Europe and even within countries social enterprise refers to a wide range of institutions. Therefore, several models have been developed to classify social enterprises. Two conceptual approaches appear to have a relatively greater use and acceptance in Europe, namely the one brought forward by the last UK government (referred to above) and the other led by EMES (European Research Network) across Europe. In the former case, the government put forward their own definition of social enterprise and carried out a first inventory of these organisations. While this can be considered a 'top-down' approach, the EMES approach corresponds to a more 'inductive' approach, i.e. a systematisation of a social and economic phenomenon already happening 'on the ground'. EMES defines social enterprise as:

> (...) not-for-profit private organisations providing goods or services directly related to their explicit aim to benefit the community. They rely on a collective dynamics involving various types of stakeholders in their governing bodies, they place a high value on their autonomy and they bear economic risks linked to their activity. (Defourny & Nyssens, 2008)

Furthermore, EMES has identified a set of nine indicators (see Table 2) that describe a social enterprise. These indicators can be distinguished according to their 'social' or 'economic' emphasis[1]:

Both definitions refer to the two main dimensions and/or objectives of social enterprises, namely the 'social' and the 'economic' (or commercial). This duality is further formulated by some authors in terms of the tension or as a continuum between both dimensions. Indeed, the 'entrepreneurial' element of this concept implies, as said earlier, a quest for efficiency and innovation that requires by definition a technical–economical

Table 2. Indicators describing social enterprise according to the EMES network.

Economic indicators	Social indicators
• A continuous activity producing goods and/or services.	• An explicit aim to benefit the community.
• A high degree of autonomy.	• An initiative launched by a group of citizens.
• Significant level of economic risk.	• A decision-making power not based on capital ownership.
• A minimum amount of paid work.	• A participatory nature, which involves various parties affected by the activity.
• A limited profit distribution.	

Source: Authors' elaboration on the basis of Defourny (2009) and WISE (www.wiseproject.eu).

rationale, which sometimes seems to clash or stand in competition with their social objectives. Moreover, a number of scholars studying not-for-profit organisations (including social enterprise) have put forward the idea of a 'continuum of practice' ranging from 'social' to 'economic' objectives, along which these organisations would hold different positions (Crossan & Til, 2009; Stull, 2003). Furthermore, some highlight the importance of the internal and external influences affecting these organisations, such as the market and governance structures (Hasenfeld & Gidron, 2005; Marwell & McInerney, 2003; Stull, 2003).

Crossan's study of social enterprises[2] in Ireland suggested that:

[t]he majority of social economy practitioners (…) agreed with the reality that social enterprises operate along a spectrum of activity, under the influence of many factors including the need to be socially driven and economically sustainable (Crossan, 2007). At the very basis level the continuum can be described as having one end of the scale where the focus for the social economy enterprise will be to have a more social and less economic focus, potentially only operating to generate enough income to survive. At the other end of the scale, the social economy enterprises will operate and present themselves as businesses that aim to maximise profits to fund underlying social objectives (SEL 2002). (Crossan & Til, 2009, p. 10)

Furthermore, Crossan proposes a conceptual classification framework called 'The Social Economic Continuum' (Figure 1) applicable to Northern Ireland and the Republic of Ireland. This continuum illustrates the location of not-for-profit organisations in relation to commercial enterprises that have a level of social focus, which may impact the management decisions and practices. As Crossan & Til explains, 'The model describes the Continuum from social activity to economic activity, by moving form the public/government sector, through the not-for-profit sector and into the for-profit sector' (2009, p. 10). To classify organisations along the Social Economic Continuum, Crossan devised a 'Theoretical Measurement Framework Model' on the basis of indicators she found to be used in the literature to classify and characterise social purpose aims and objectives, regardless of the sector of the organisation. These indicators can be divided into three layers: (1) the *descriptor variables* (e.g. legal structure, business model, staffing, trading activities, etc.)[3]; (2) *motivator variables* (e.g. purpose of the organisation or their social

Not-for-profit social		Social commercial			For-profit commercial social		Commercial
			Community	Businesses			
				Housing assoc			Soletrader
				Mutuals			
		Charities		Credit unions		Shareholder	SME
Quangos Community		with co.					
Govnt & voluntary Charities		Ltd G	Social firms	Co-operatives	Family owned	Ethical	Co ltd G

(Outwardly social) ⟶ Inwardly social) | (Ourwardly social ⟶ Inwardly social)

Social activity | Economic activity ⟶

Figure 1. The conceptual classification framework: the social economic continuum for NI and Northern Ireland and Republic of Ireland. *Source:* Crossan, 2007.

aims and objectives); and (3) *behaviour variables* (i.e. how the organisation actually performs in terms of meeting their social aims and objectives). Interestingly, the behaviour variables act as the 'true test' to decide the position of any organisation along the Social Economic Continuum. (Crossan & Til, 2009, p. 8).

The continuum of Crossan implies that social enterprises are hybrid institutions, combining values and activities from state, market and community organisations to different extents. Therefore, it is also very much in line with the work by Evers (2005) on the welfare mix discussed above and the model developed by Brandsen *et al.* (2005), based on Pestoff (1992) and Zijderveld (1999), to classify hybrid organisations. Within their conceptualisation, the 'third sector' is a hybrid domain amidst the three ideal-typical or 'pure' domains of society (i.e. organisations in this sector emerge as hybrid types between the pure actors we know as bureaucracies, enterprises, and families or clans), as is reflected on in the triangle in Figure 2. Brandsen *et al.* (2005) point out that hybrid organisations can be found at different positions within this triangle and also acknowledge that 'borders' between various types of organisations are fuzzy. One limitation of this model may be that the third sector is not seen as a domain in its own right but rather as a tension field between the state, market and community (Buckingham, 2010). In this respect it differs from recent work on hybridity such as that by Billis (2010) which includes the third sector as a domain, referring to a pure organisational form of the membership organisation.

The notion of hybridity has been at the centre of debates about social enterprise in England and Billis (2010) has identified nine 'zones of hybridity' at the boundaries between the public, private and third sectors. He argues that hybrid organisations can be divided on the basis of their 'principal ownership', which will usually lie within one of the three sectors. Hybridity may be shallow or entrenched and may be organic (as in the case of 'traditional' housing associations with civil society roots) or enacted (as in the case of stock transfer housing associations). In the same volume, Aitken argues that social enterprises provide 'exemplars of the hybrid form ... since they intertwine within a single organisation different components and rationales of state, market and civil society' (2010, p. 153). He goes on to conclude that as hybridity becomes increasingly common it will be important for policy makers to understand how social enterprises manage tensions between competing principles.

Organic paradigm shift to entrenched HAs in NI

want to examine tensions between competing principles at a corporate governance / board level.

21

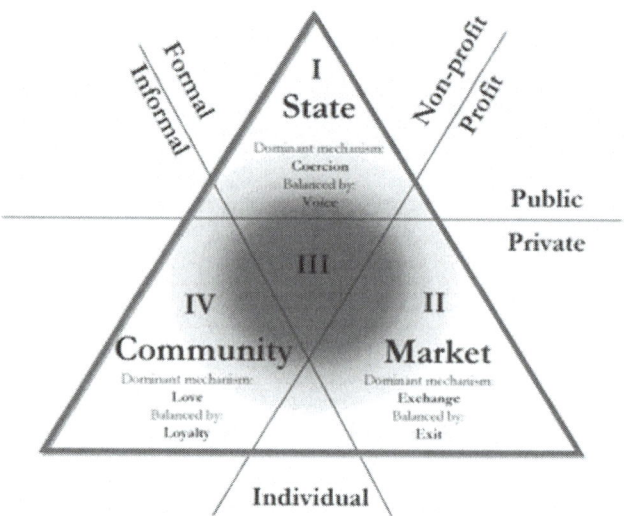

Figure 2. Position of social enterprises between state, market and community organisations. *Source:* Brandsen *et al.* (2005), based on Zijderveld (1999) and Pestoff (1992).

Mullins and Pawson explore the application of hybridity to English and Dutch housing associations identifying a number of common elements as both sectors have become entrenched hybrids with 'large, bureaucratised organisations with paid staff becoming the principal owners of strategy as well as operations' (2010, p. 213). They suggest that the two sectors may fall into rather different zones of hybridity: 'as the Dutch sector has become a private/third sector hybrid it has been increasingly concerned to protect itself from the "profits in disguise" critique by cultivating a social entrepreneurial identity and developing explicit social benefits' (2010, pp. 213–214). Meanwhile:

> the English sector seems closest to the public/private/third sector zone nowadays and this explains their robust defence against the dangers of simply becoming agents of public policy and generally positive presentation (at least before the credit crunch) of its leverage of private funds and know-how. (2010, p. 214)

Work on hybridity is helpful in indicating the competing principles that may affect organisational positioning among social enterprises. However, it is important to understand how these principles are applied in organisational strategies and decisions, to balance often competing social and commercial principles. Stull's work on decision-making within hybrid organisations (such as social enterprises) sees this as a day-to-day dynamic process involving decision makers such as CEO's blending 'traditional' with 'innovative' approaches with an aim to balance the (core) social mission with constant drivers towards maximising economic efficiency. Hence, in the management of every social enterprise decisions that have to be made will vary between bi-polar characteristics in a number of aspects of the company's operations, ranging from mission types to funding sources, etc. (see Figure 3).

In Stull's continuum of practice, the 'traditional management approach' is defined as having:

(...) an intense commitment to the mission and tradition (values and practices) of the organisation and a viewpoint of 'that's the way we have always done it'. Mission and traditionality is viewed as a constant, inelastic dimension of the organisation; it is to be reinterpreted or reoriented only when fundamental change dictates the need to transform the organisation (Salipante & Golden-Biddle, 1995). The organisation is highly focused on internal processes and program content versus specific outcomes as they relate to the organisation and market, and is often limited or defined by its external funding sources and organisational resources. (Stull, 2003, pp. 14–15)

[handwritten margin note: what dictates? the change? Reduction in public expenditure]

The other end of the continuum, representing 'social entrepreneurship', is defined by Stull as an opportunity-oriented approach:

... that attempts to integrate both mission and market views. Mission is viewed as elastic in nature, able to be constantly evaluated and reinterpreted to meet the changing needs of the market and the organisation. The organisation is positioned based on market demands with a specific or desired outcome that is directly related to the mission. Market opportunities are not defined or judged based on availability of funding resources, nor is the organisation constrained by its traditional funding sources. (Stull, 2003, p. 15)

Teasdale (2009a) also uses the social-economic continuum and adds a second dimension to classify social enterprises. He suggests that social enterprises can be categorised into four main groups on the basis of two scales: whether the main motivation is social or economic and whether the main organisation is at the individual or collective level (see Figure 4). Social businesses are the dominant group numerically, but co-operatives and community enterprises are closer to what policy makers may have had in mind in promoting social enterprise.

In another paper, Teasdale (2009a) highlights the ways in which social enterprises actively manage their identities to 'exhibit multiple faces to different stakeholder in order to access resources' (p. 1). His application of Goffmann's concept of impression management to a case study organisation to a theatre company in the Kurdish refuges community highlights the impact of hybridity not just on the different faces the

Traditional management characteristics	Social entrepreneurship characteristics
Mission & tradition driven	Mission & market driven
Inelastic mission	Elastic mission
Client & internal process orientation	External market and outcomes orientation
Defined by funding sources	Unconstrained by funding sources
Administrative orientation	Entrepreneurial orientation

Figure 3. Stull's continuum of practice with traditional management and social entrepreneurship as relative extremes. *Source:* Stull (2003).

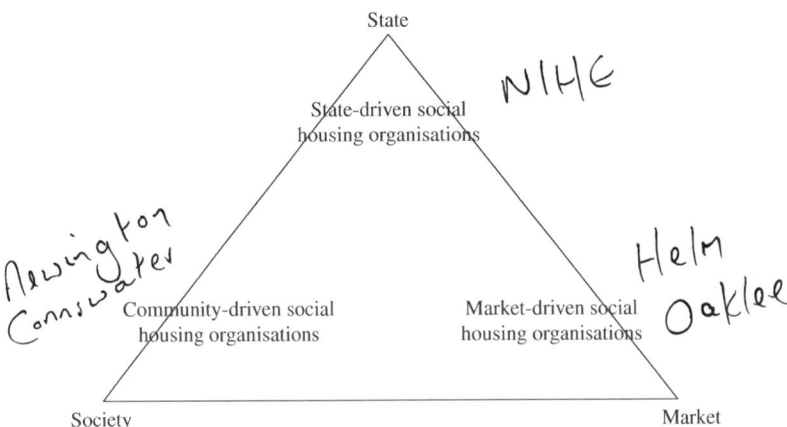

Figure 4. Positions of social housing organisations between state, market and society. *Source: Developed from Czischke & Gruis (2007) and Brandsen et al. (2005).*

organisation consciously presents to outsiders, but also how internal stakeholders see the social enterprise in different ways. There are clear resonances of this in the English housing association sector where the trade body's mission 'in business for neighbourhoods' offers endless possibilities of accentuating either the commercial ethos symbolised by the mantra 'in business' or the socially responsible and altruistic alter ego that they are only there 'for neighbourhoods'.

In summary, social enterprise can be viewed from an institutional perspective and from a behavioural perspective. From an institutional point of view, it is concluded that social enterprises can take different formal, legal shapes, but have in common that they are (to different degrees) hybrid institutions that deal with competing drivers stemming from public, market and community forces and values. These drivers occur in varying combinations, sometimes reflecting the origins of these organisations (e.g. Billis' distinction between organic and enacted hybridity) and affecting the motivations of the key actors involved. From a behavioural perspective, social enterprises have in common that they (again to different extents) adopt entrepreneurial strategies to fulfil their objectives (versus traditional 'bureaucratic' or 'management-oriented' behaviour); conflicts between principles are played out in organisational strategies and day-to-day decisions.

(Key to thesis – day to day decisions at a governance level.

A Classification Framework for Social Enterprise in Housing Organisations

According to the general definitions and characteristics of social enterprise discussed in the previous section, various social housing organisations can be brought under the general heading of social enterprises. Nevertheless, it is well known that there is a great variety in the nature and organisation of social housing providers between and even within countries. Therefore, to study and compare the characteristics of social enterprise within social housing organisations, it is useful to develop a classification model for this specific sector. In this section, we present a classification model that is built on the classification frameworks developed by scholars in the field of social enterprise in general, combined with preliminary research from the social housing sector based on work with case study companies in England, the Netherlands and Finland.

Following Crossan & Til (2009), our framework uses a distinction between descriptor values (formal institutional characteristics), motivator values (as reflected in organisation's missions and drivers) and behaviour values (as reflected in the organisations' approaches and output or activities).

For the descriptor as well as the motivator values, we use the triangle from Brandsen *et al.* (2005) and Billis's notion of zones of hybridity as a reference point. Applying these models to the discussion of the concept of social enterprise in the previous section, we conclude that:

(1) social enterprises can be placed in zones that are located in the overlaps between pure government (public), market (commercial) or community (informal) organisations; often they mix these forms in ways that reflect their origins and principal ownership;

(2) they encompass competing values and approaches from all three 'pure' institutional contexts;

(3) values and approaches from one angle may be relatively dominant over values and approaches from other angles (and the relative dominance may change over time);

(4) they can have a legal structure that is either (semi)governmental, (semi)private, (semi)informal or any other hybrid structure;

(5) (this implies) social enterprises may encompass non-profit and/or not-for-profit[4] as well as limited-profit organisations;

(6) nevertheless, if the values and approaches from one corner are completely dominant over values and approaches from other corners, the organisation cannot be classified as a social enterprise.

To translate the 'position' that a social housing organisation occupies within the triangle according to *descriptor* variables, the following characteristics can be used:

Mike Gaskell

(1) the legal structure (public, private, informal, other, including hybrid legal forms such as community interest companies in the UK);

(2) the profit objective (non-profit, not-for-profit, limited);

(3) governance (in which institutes or people are formally responsible for policy-making within the housing organisations: state, tenants, societal and/or commercial stakeholders or shareholders).

As we can see in Table 3, there is a wide variety of social housing organisations in Europe according to their formal, descriptive characteristics. There is a mix of (formally) government-owned social landlords, such as municipal housing companies, tenant-owned cooperatives as well as private housing associations, some with shareholders other than tenants (such as the German housing companies), some without (e.g. Dutch housing associations). Some of these organisations operate explicitly under a non-profit policy (e.g. Danish housing associations). Others operate on a not-for-profit basis, but may accumulate financial reserves over time (e.g. Dutch housing associations) or on a limited profit basis (such as the Austrian housing companies). The (formal) governance mostly coincides with the legal structure, although exceptions should be noted. For example, housing management within Danish housing associations is largely controlled by tenants

Table 3. Type of social housing providers in the EU-15 ('old member states').

	Under state control/ State owned	Public or publicly controlled companies	Not-for-profit social housing companies	Social housing companies
Germany	No	• Municipal companies • Public companies (Bund et Lander)	• Co-operatives	Yes
Austria	Yes	No	• Associations and companies	No
Belgium	No	• Municipal companies	No	No
Denmark	No	• Municipal associations	• Associations made of Independent management units	No
Spain	Yes	• Public companies	• Co-operatives	Yes
Finland	Yes	• Municipal associations	• Co-operatives • Associations • Co-operatives	Yes
France	No	• Public bodies • Companies of mixed economy	• No for profit companies	Yes
Greece	Yes	• OEK (Workers' Housing Organisation, Greece)	• Co-operative companies	No
Ireland	Yes	No	• Housing associations/Co-operatives	No
Italy	No	• Local public housing companies • Fond du logement à cout modéré (Low-cost housing Fund)	• Co-operatives	Yes
Luxembourg	Yes	• Société nationale des habitations a bon marché (National company of low-cost housing)	No	No
The Netherlands	No	No	• Enterprises (corporaties)	No
Portugal	Yes	• Public body	• Co-operatives • Charities	Yes
United Kingdom	Yes	Yes (ALMOs: Arms Length Management Organisations)	• Approved social landlords	Yes (marginal)
Sweden	No	• Municipal companies	• Co-operative companies	No

Source: Heino *et al.* (2007).

but is also subject to detailed government regulation (Gruis & Nieboer, 2004). Dutch housing associations, on the other hand, have a large administrative and financial independence, but are actively seeking ways to incorporate societal stakeholders in their decision-making.

For the *motivator variables*, according to Crossan, one must look at the purpose and social aims of the organisations. In the case of social housing organisations these are reflected in their missions and objectives and the motivations underpinning the missions. Following Branden's triangle, a distinction can be made between:

(1) *state-driven housing organisations* whose missions are relatively strongly influenced by state policies, regulations and finance;
(2) *market-driven housing organisations* whose missions are relatively strongly influenced by (general) housing market demand and financial-economic opportunities on the housing market and depending relatively much on private finance;
(3) *community-driven housing organisations* whose missions are relatively strongly influenced by preferences and financial means of their current tenants.

In this respect, the findings of a study of six social housing organisations from different European countries by Heino *et al.* (2007, p. 78) are worth noting:

> Overall, the core task of all companies can be broadly formulated as to provide rental dwellings for households who have difficulty in accessing a home in the market, either on the basis of income or other type of disability … According to each national context, the company's mission is defined by law as a public service obligation to match specific policy objectives. As can be seen in Table 1, all companies respond to a specific mission related to providing housing for people in need, the latter being defined by different criteria in each country. However, the degree of independency to define their company's mission and core task varies significantly between them. While in France and Italy the company's mission and core task arise from a legal framework that defines what a public service obligation is, in the Netherlands—and to some extent in the UK— the companies enjoy greater freedom to define these. For [Company E (an English company)], for example, the interviewees describe the company's mission as a 'moral obligation'. In the case of [Company N (a Dutch company)], there is a sense of self-imposed social obligation towards the community and society in the context of increasing withdrawal of the state from a series of welfare areas.'

To analyse the characteristics of social housing organisations according to Crossan's *behavioural variables*, one can look at the actual activities and strategic decisions that they perform to fulfil their missions. Several studies have indicated social housing organisations across Europe are following diversification strategies to different extents (e.g. Brandsen *et al.*, 2006; Heino *et al.*, 2007; Mullins *et al.*, 2001; Walker, 2000). The survey conducted among 42 social housing organisations spread across 12 different EU states by Heino *et al.* (2007) indicates there is a considerable degree of dynamics in the activities of social landlords (see Figure 5). The importance of the motivator variable in distinguishing between diversification undertaken simply to earn income and that undertaken to also better meet community needs is clear. However, these motivations can be difficult to

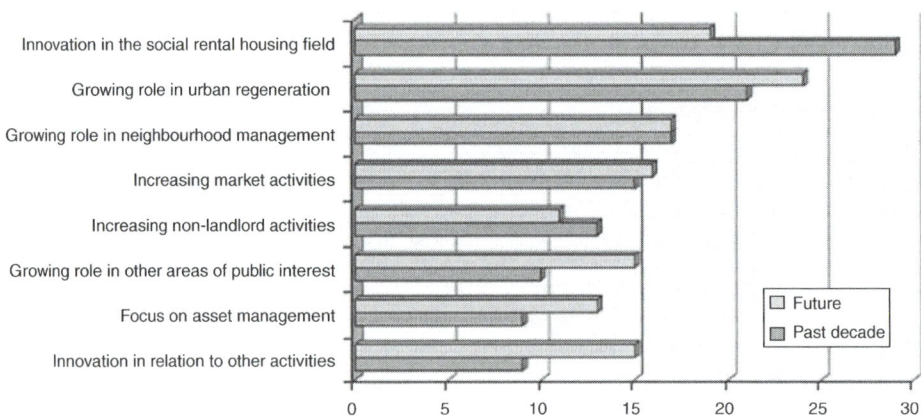

Figure 5. Recent and expected changes in the activities of 42 social housing organisations in Europe. *Source:* Heino et al. (2007).

unpack in the context of specific decisions, which are likely to have multiple rationales; the easiest motivational impacts probably involve commercial opportunities forgone because of conflicts with values; e.g. decision not to expand into activities that involve low wage competition and conflict with values about how staff is treated (Mullins & Riseborough, 2000).

To facilitate classification according to behavioural values in the context of social housing organisations, we build on the work of Gruis (2005, 2008) to typify the extent and nature of the diversification of the actual activities (output) of the housing organisations. Gruis (2008) uses Miles and Snow's distinction between 'Prospectors' and 'Defenders', which refers to types of strategic orientations that can be found among commercial enterprises and which can also be related to Stull's bi-polar typology. As he explains, some social landlords are focussed mainly on performing their core, traditional (public) task of providing decent, affordable housing for those in need, while other social landlords provide additional 'public' services (such as care and welfare services) or are active within commercial segments of the housing market as well (by providing expensive rental and owner-occupied dwellings).

When applying the 'Defenders–Prospectors' distinction, Gruis found that Defenders among social housing organisations focus on traditional activities, i.e. building and managing social rented housing for lower-income households and are not focussed on innovations outside this area. Their main objective is to perform their current task in a good manner and (possibly) to increase efficiency. In a sense, Defenders are following Walker's (2000) scenario of focussing on their core, property-driven, task. This would fit with Stull's 'traditional management characteristics', as seen above. On the other hand, Prospectors among social housing organisations are focussed on innovations. They seek new opportunities within their current field of operations as well as to expand their field of operations. They are focussed on increasing their effectiveness (in various areas of their operations), rather than on maximising efficiency. Prospectors can be regarded as an analogue to Stull's 'social entrepreneurs'.

However, following the classification of Miles & Snow (1978), Prospectors can be further divided into social housing organisations that primarily seek to enlarge their activities in the *(semi) public domain* and social housing organisations that focus on expansion of *market* activities. Prospectors in the first group follow Walker's (2000) scenario of developing the housing plus services, while Prospectors in the second group maintain their primary orientation on property, but become more active in commercial segments of the housing market as well (higher rent and sale). In addition, a third group of Prospectors combines diversification within the public domain with expansion of commercial activities. Figure 6 summarises the types of social housing providers according to their diversification strategy.

Within the classification developed above, it seems likely that there is a relationship between the descriptive, motivational and behavioural characteristics, which would open up the possibility of a single classification. For example, a study by Mullins & Sacranie (2009) to understand the motivating drivers of 'community investment activities in English housing associations pointed at a relationship between descriptor variables and motivator variables. They concluded that the ways in which organisations construct, prioritise and measure their community investment activities will depend on whether their principal ownership is primarily market-driven, in which case CSR (Corporate Social Responsibility) will be driven by their corporate business strategy in the same way as a private company. If, on the other hand, they are more society-driven (e.g. neighbourhood focused housing associations) then community investment activities will emerge from locally based relationships between staff and residents and success will be judged through local stakeholders' feedback. Nevertheless, evidence from earlier explorative research (Gruis, 2008; Heino *et al.*, 2007) suggests that there is not a 100 per cent relationship between these three elements of Crossan's framework within organisations. For example, some housing organisations that formally have a relatively large freedom in determining their own policies, to actively pursue profits with market activities and have adopted an 'entrepreneurial' business mission have been found to be having a traditional scope of activities. Therefore, in line with Crossan, we argue one should look at all three of the variables individually to classify the nature of social enterprise within housing organisations, although it remains interesting to see what kind of relationships exist between the three variables.

Conceptual organisational archetypes of housing associations.

	Defender		
Social orientation	**Social-housing manager** emphasis on fulfilling traditional tasks, affordability and tenant satisfaction	**Social-housing investor** emphasis on traditional activities, business efficiency and yielding profit via going-concern approach.	Commercial orientation
	Societal innovator emphasis on broad range of activities, continuous renewal of its products and services, and social return.	**Societal real-estate investor** emphasis on continuous renewal of its real estate portfolio and on financial return	
	Prospector		

Figure 6. Types of social housing providers according to diversification strategy. *Source:* Gruis, 2008 (adapted by Nieboer and Gruis, 2011).

Conclusion

In this paper we have explored the concept of social enterprise and translated this concept into social housing organisations in the shape of a classification framework. Although in some countries, social enterprise seems to refer mostly to 'third sector' or 'community' organisations in other countries and more recent studies social enterprise refers to a wide range of organisations that have a social mission and are not 'purely' commercial, state (bureaucratic) or (restricted) community-based organisations. According to these definitions, in addition to their social purpose, social enterprises are characterised by hybrid formal institutional characteristics, motivations and activities. Within their hybrid nature, a wide variety of organisations exists, which are classified in various studies on continua between having a relative emphasis on social or commercial objectives and activities (e.g. Crossan & Til, 2009) and bureaucratic, democratic or entrepreneurial mechanisms for management (e.g. Brandsen *et al.*, 2005; Stull, 2003; Teasdale, 2009a, 2009b).

According to the general definitions and characteristics of social enterprise, various social housing organisations can be brought under the general heading of social enterprises. Nevertheless, there is a great variety within this specific sector of social enterprises as well. Following Crossan, we have argued that social housing organisations can be classified according to descriptive variables (formal institutional characteristics), motivator variables (missions and drivers) and behavioural variables (nature and range of activities).

The classification allows comparison between different organisations operating in different contexts, which share a similar core task, namely to provide housing for those who cannot afford it in the market and/or who have special needs. This can be useful for scientific, political and practical purposes. Within comparative housing studies, it can be used as a basis for comparing similarities and differences between social housing organisations within and between countries at a much deeper level than the 'traditional' comparisons on the basis of tenure or other formal organisational characteristics. Thereby it can also generate interesting information for policy makers to see to what extent different political, economical and institutional contexts lead to different behaviour by social housing providers. Furthermore, the model may also be useful for professionals, to assess to what extent their social housing organisation is being consistent with regards to what they say they (want to) do and what they actually do. However, for all purposes, more measurable information (indicators) has to be obtained by applying the model to individual housing organisations, which will be a main task for further research. For example, this classification is currently being tested in an ongoing PhD research (Czischke 2010) comparing social housing organisations in three different European countries. We expect that the findings of this and other current research will contribute to test and improve the proposed classification.

Acknowledgements

We would like to thank colleagues in the ENHR Working Group 'Social Housing: Institutions, Organisations and Governance' and at our two research centres (MOVe at TU Delft and TSRC at the University of Birmingham) for their insightful comments on various drafts of this paper and to the three anonymous referees from Housing Studies.

Notes

[1] See WISE project (Work Integration Social Enterprises) at www.wiseproject.eu

[2] As can be noted from the selected quotation, the author uses the terms 'social enterprise' and 'social economy enterprise' either interchangeably, or taking one as a sub-type of the other. As we have explained earlier in this paper, this use of the terms often happens in this field of research.

[3] Care is required here since, for example, in the UK companies limited by guarantee are a common legal form for charities, while most family businesses are also SMEs.

[4] The distinction between 'non-profit' and 'not-for-profit' is subtle. In fact, in academia, policy and practice, both terms are often used interchangeably. 'Not-for-profit' characterises the non-distributing nature of these organisations. However, in practice different third sector organisations adopt one term or the other following their own concept of non-distribution and their national specific legal and linguistic traditions. Hence, in our classification we have decided to include both the terms to encompass the wide variety of self-definitions that can be found in practice as well as in the literature in this field. Austrian (social) housing cooperatives have the specific legal status of 'limited profit' organisations.

References

Aitken, M. (2010) Social enterprises: challenges from the field, in: D. Billis (Ed.) *Hybrid Organizations and the Third Sector: Challenges for Practice, Theory and Policy*, Chapter 8, pp. 153–174 (London: Palgrave Macmillan).

Alcock, P. (2010) *Partnership and Mainstreaming: Voluntary Action Under New Labour*. Third Sector Research Centre Working Paper 32 (Birmingham: University of Birmingham).

Bidet, E. (1997) *L'économie sociale* (Le Monde-Editions). Paris, Le Monde-Éditions, coll. « Le Monde Poche », Marabout n° 8663.

Billis, D. (2010) Towards a theory of hybrid organisations, in: D. Billis (Ed.) *Hybrid Organizations and the Third Sector. Challenges for Practice, Theory and Policy*, Chapter 3, pp. 46–69 (London: Palgrave Macmillan).

Boelhouwer, P. (1999) International comparison of social housing management in Western Europe, *Netherlands Journal of Housing and the Built Environment*, 14, pp. 225–240.

Brandsen, T., van de Donk, W. & Putters, K. (2005) Griffins or chameleons? Hybridity as a permanent and inevitable characteristic of the Third Sector, *International Journal of Public Administration*, 28, pp. 749–765.

Brandsen, T., Franell, R. & Cardoso Ribeiro, T. (2006) *Housing Association Diversification in Europe; Profiles Portfolios and Strategies* (Withley Village: The REX Group).

Brouard, F. (2007) Réflexions sur l'entrepreneuriat social, in: Z. W. Todorovic (Ed.) *Entrepreneurship and Family Business: Proceedings of the Annual Conference of the Administrative Sciences Association of Canada*, June 3–5, Ottawa, Ontario, volume 28, Division 21.

Buckingham, H. (2010) *Capturing Diversity: A Typology of Third Sector Organisations' Responses to Contracting Based on Empirical Evidence From Homelessness Services*. TSRC Working Paper 41 (Birmingham).

Crossan, D. (2007) Towards a classification framework for not for profit organisations, PhD dissertation thesis, School of International Business, University of Ulster, 2007.

Crossan, D. & Til, J. V. (2009) Towards a Classification Framework for Not-For-Profit Organisations: The Importance of Measurement Indicators, EMES Selected Conference Paper Series. EMES.

Czischke, D. (2010) *Understanding Organisational Development of Social Housing Organisations in the European Union*. Second PhD report, Delft University of Technology.

Czischke, D. & Gruis, V. (2007) Managing social rental housing in the EU in a changing policy environment: Towards a comparative study. Paper presented at the Workshop Comparative Housing Policy, Dublin, Centre for Housing Policy, April 20–21.

de Boer, N. (1999) *Maatschappelijk ondernemen in de gezondheidszorg; wat en hoe?* [Social Enterprise in Health Care; What and How?] (Amsterdam: De Balie).

Defourny, J. (2009) Concepts and realities of social enterprise: A European perspective, *Collegium* (38, Spring), pp. 73–98.

Defourny, J. & Nyssens, M. (2008) *Social Enterprise in Europe: Recent Trends and Developments*. EMES Working Paper series, no. 08/01, Liège.

de Kam, G. (2003) Op grond van betekenis: over maatschappelijk ondernemen met grond en locaties [On the ground of meaning: on social enterprise with land and locations]. Available at http://www.ru.nl/contents/pages/11730/redegeorgedekam.pdf

Delors, J. & Baudin, J. (1979) Pour la création d'un troisième secteur coexistant avec celui de l'économie de marché et celui des administrations', *Problèmes économiques*, 1616, pp. 20–24.

DTI (2002) *Social Enterprise Strategy for Success* (London: Department of Trade and Industry (DTI)).

Evers, A. (2005) Mixed welfare systems and hybrid organisations: Changes in the governance and provision of social services, *International Journal of Public Administration*, 28(9), pp. 737–748.

Evers, A. & Laville, J. -L. (2004) Defining the third sector in Europe, in: A. Evers & J. -L. Laville (Eds) *The Third Sector in Europe* (Cheltenham: Edward Elgar).

Gilmour, T. (2009) Network power: An international study of strengthening housing association capacity, PhD thesis, University of Sydney.

Gruis, V. (2005) Bedrijfsstijlen woningcorporaties; hulpmiddel bij het invullen van het maatschappelijk ondernemerschap [Organisational archetypes housing associations; support for shaping social enterprise], *BuildingBusiness*, (7), pp. 54–57.

Gruis, V. (2008) Organisational archetypes for Dutch housing associations, *Environment and Planning C: Government and Policy*, 26(6), pp. 1077–1092.

Gruis, V. (2010) De woningcorporatie als katalysator; over de missie en organisatie van maatschappelijk ondernemende woningcorporaties [The housing association as catalyst; on the mission and organisation of socially entrepreneurial housing associations], *BuildingBusiness*, Amsterdam.

Gruis, V. & Nieboer, N. (Eds) (2004) *Asset Management in the Social Rented Sector; Policy and Practice in Europe and Australia* (Dordrecht: Springer/Kluwer Academic Publishers).

Hasenfeld, Y. & Gidron, B. (2005, September) Understanding multi-purpose hybrid voluntary organizations: The contributions of theories on civil society, social movements and non-profit organizations, *Journal of Civil Society*, 1(2), pp. 97–112.

Heino, J., Czischke, D. & Nikolova, M. (2007) *Managing Social Rental Housing in the European Union: Experiences and Innovative Approaches* (Brussels: CECODHAS European Social Housing Observatory & VVO-PLC).

Kerlin, J. A. (2006) Social enterprise in the United States and Europe: Understanding and learning from the differences, *Voluntas*, 17(3), pp. 246–262.

Lyon, F. & Sepulveda, L. (2009) Mapping social enterprise: Past approaches, challenges and future directions, *Social Enterprise Journal*, 5(1), pp. 83–94.

Marwell, N. P. & McInerney, P. B. (2003) The nonprofit/for-profit continuum: Making forprofit markets from nonprofit activities. Paper presented at the ARNOVA 2003 Conference, Denver, Colorado.

Mertens, S. (1999) Nonprofit organisations and social economy: Two ways of understanding the third sector, *Anals of Public and Cooperative Economics*, 70(3), pp. 501–520.

Miles, R. E. & Snow, C. C. (1978) *Organizational Strategy, Structure, and Process* (New York: McGraw-Hill).

Mullins, D. & Pawson, H. (2010) Housing associations: Agents of policy or profits in disguise? in: D. Billis (Ed.) *Hybrid Organizations and the Third Sector: Challenges for Practice, Theory and Policy*, Chapter 10, pp. 197–218 (Basingstoke: Palgrave Macmillan).

Mullins, D. & Riseborough, M. (2000) What are housing associations becoming? Working Paper No. 9, CURS (Birmingham: The University of Birmingham).

Mullins, D. & Sacranie, H. (2009) *Corporate Social Responsibility and the Transformation of Social Housing Organisations: Some Puzzles and Some New Directions* (Prague: ENHR).

Mullins, D., Latto, S., Hall, S. & Srbljanin, A. (2001) *Mapping Diversity: Registered Social Landlords, Diversity and Regulation in the West Midlands Birmingham*. Working Paper No. 10, CURS (The University of Birmingham: School of Public Policy).

NHF (2010) *Annual Conference and Social Housing Exhibition 2010* (London: National Housing Federation).

Nieboer, N. & Gruis, V. (2011) Shifting back in the Dutch social housing sector. Paper presented at the 23rd ENHR 2011 conference, 2–8 July, Toulouse, France.

NTMO (2003) De waarde van de maatschappelijke onderneming geborgd [The value of the social enterprise secured]. NTMO, Hilversum.

Office for the Third Sector (2006) *Social Enterprise Action Plan* (Scaling New Heights) (London: Cabinet Office).

Pawson, H. & Mullins, D. (2010) *After Council Housing: Britain's New Social Landlords* (Basingstoke: Palgrave).

Pestoff, V. A. (1992) Third Sector and co-operative services: An alternative to privatisation, *Journal of Consumer Policy*, 15, pp. 21–45.

Priemus, H., Dieleman, F. & Clapham, D. (1999) Current developments in social housing management, *Netherlands Journal of Housing and the Built Environment*, 14(3), pp. 211–224.

Salipante, P. F. & Golden-Biddle, K. (1995) Managing traditionally and strategic change in nonprofit organizations, *Nonprofit Management & Leadership*, 6(1), 3–20. quoted in Stull, M. (2009) op cit.

SER (2005) *Ondernemerschap voor de publieke zaak* [Entrepreneurship for the public cause] (Den Haag: SER).

Social Enterprise London (2002) *Social Enterprise: A Strategy for Success* (London: SEL).

Stull, M. G. (2003) Balancing the dynamic tension between traditional nonprofit management and social entrepreneurship. Paper presented at the ARNOVA 2003 Conference, Denver, Colorado (USA).

Teasdale, S. (2009a) *Can Social Enterprise Address Social Exclusion? Evidence for an Inner City Community.* Working Paper No. 3 (Third Sector Research Centre).

Teasdale, S. (2009b) *The Contradictory Faces of Social Enterprise: Impression Management as Social (Entrepreneurial) Behaviour.* Working Paper No. 23 (Third Sector Research Centre).

van Dijk, G., Klep, L. M. F., van der Maden, R., Duit, I. G. A. & van Boekel, P. (2002) *De woningcorporatie als moderne maatschappelijke onderneming* [The housing association as modern social enterprise] (Assen: Koninklijke Van Gorcum).

Vulperhorst, L. (1999) *De kern van de zaak; maatschappelijk ondernemen door woningcorporaties* [The core of the matter: social enterprise by housing associations] (Utrecht: Andersson Elffer Felix).

Walker, R. M. (2000) The Changing management of social housing: The impact of externalisation and managerialisation, *Housing Studies*, 15, pp. 281–299.

Young, D. (2001) Social enterprise in the United States: Alternate identities and forms. Paper presented at the International EMES Conference, Trento, Italy, December.

Young, D. (2003) Social enterprise in community and economic development in the United States: Organizational identity, corporate form and entrepreneurial motivation. Paper presented at the International Workshop on Modern Entrepreneurship, Regional Development and Policy: Dynamic and Evolutionary Perspectives, Tinbergen Institute, Amsterdam.

Zandstra, A. & Rohde, W. (2002) *Maatschappelijk ondernemers gezocht* [Social entrepreneurs wanted] (Amsterdam: RIGO Research and Advies BV).

Zijderveld, A. C. (1999) *The Waning of the Welfare State: The End of Comprehensive State Succor* (New Brunswick, NJ: Transaction Publishers).

The Quadruple Bottom Line and Nonprofit Housing Organizations in the United States

RACHEL G. BRATT

Department of Urban and Environmental Policy and Planning, Tufts University, Medford, USA

ABSTRACT *The work of US nonprofit housing organizations can be viewed as involving a commitment to meet the Quadruple Bottom Line—the financial demands of developing and maintaining affordable housing while serving resident groups and neighborhoods, in an environmentally responsible manner. Nonprofit organizations may be categorized into three major groups, based on their primary areas of concern—'people', 'places' and 'projects'. This article outlines the components and approximate size of the US social housing sector and presents examples of how housing nonprofits have, both historically and currently, evolved to incorporate multiple roles. With declines in federal funding for affordable housing, nonprofits have become increasingly hybrid in their operations. Examples are given regarding how nonprofits attempt to mediate the private market; how the various components of the Quadruple Bottom Line often compete with one another; and how hybridity of the nonprofit social housing sector creates additional challenges for these groups. The final section presents policy directions for supporting nonprofits.*

Introduction

Nonprofit housing organizations have been operating in the US since the early 20th century. With disparate missions, orientations and histories, they were not, at that time, viewed as a distinct sector. In addition, their work was modest in scope, with few units produced. Over the past 50 years, the US's nonprofit housing organizations and entities that support their work have 'come of age'. Nevertheless, with only about 5 per cent of US housing owned by public or nonprofit entities, the size of the US 'social housing' stock is far smaller than that of many European countries.

While a number of US researchers have explored the operations, challenges and opportunities presented by nonprofit housing, academics in other parts of the world, notably Europe, have been more engaged developing theoretical frameworks as a way of advancing our knowledge of the social housing sector. Connected to this is the observation that the language that Europeans use to describe what US academics typically refer to as 'nonprofit housing' or 'third sector housing' is far more nuanced. Phrases such as 'social housing', 'social enterprise' and 'hybridity' are essentially absent from US discourse.

Social enterprise is an overarching concept that includes both social housing organizations and other types of not-for-profit entities. A simple, straightforward definition of social enterprise is: 'a business with primarily social objectives whose surpluses are principally reinvested for that purpose in the business or the community, rather than being driven by the need to maximize profits for shareholders and owners' (Department of Trade and Industry, 2002, quoted in Czischke *et al.*, 2010, p. 4).

Another definition of social enterprise focuses less on the profit motive (or lack thereof) and more on the processes that such organizations use. Thus, Defourny & Nyssens (2008, p. 5) present the European Research Network (EMES) definition of social enterprise as: 'not-for-profit private organizations providing goods or services directly related to their explicit aim to benefit the community. They rely on a collective dynamic involving various types of stakeholders in their governing bodies, they place a high value on their autonomy and they bear economic risk limited to their activity'.

Others have defined nonprofit/third sector housing as being characterized by 'nonmarket relations for the production, management and ownership of land and housing' (Goetz, 1993, p. 85) that is 'price restricted and socially-oriented' (Davis, 1994, pp. 5–6). Koschinsky (1998) elaborates on these definitions by pointing out that this housing also often incorporates resident control and citizen participation.

It is likely that many, if not most, contemporary nonprofit social housing organizations are hybrids—entities that possess significant characteristics of more than one sector— public, private and third (Billis, 2010; see also Gilmour, 2009). Rubin (2000) discusses community-based development organizations as 'niche', organizations. But the essence of these groups is their hybridity, which combines the agendas of for-profit entities, government, social service agencies and community activists. While hybrid organizations likely have a clear orientation to one of the three sectors, they embrace key characteristics of at least one other. In carrying out their tasks, they may confront challenges when trying to bridge the distinctive principles and orientations of various sectors. In particular, they must mediate conflicts between the market-based operations of the private housing market and the socially oriented objectives of nonprofits.

Not all researchers see clarity emerging from these various efforts to arrive at definitions. On the basis of his deep familiarity with the literature, Teasdale has observed, 'there is still no clear understanding as to what a social enterprise is or does' (2010, p. 3). My modest effort at advancing clarity starts with the following definition, which builds on the above conceptualizations. Social housing enterprises in the US generally focus on developing or maintaining dwelling units that are affordable to a specific group of residents. Rather than profit maximization being the goal of the entity owning the housing, providing a resource for a community's long-term use is the overriding mission. In the case of homeownership units, long-term affordability is balanced against providing low-income households a modest level of equity appreciation.

Nonprofit social housing organizations in the US also are typically closely connected to the community and provide local residents opportunities for participating in decisions related to specific developments and to the general operations of the organization. Beyond

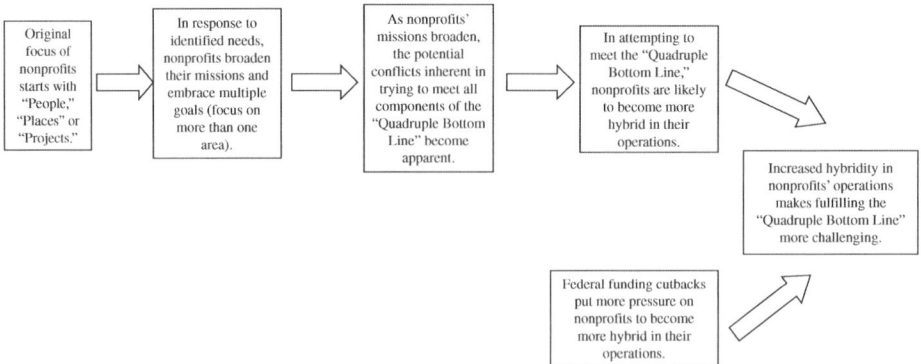

Figure 1. Schematic overview of relationships between concepts discussed.

the provision of affordable housing, most such organizations are engaged with other activities aimed at enhancing the lives of individuals and revitalizing neighborhoods. And, increasingly, they reflect the needs and priorities of the for-profit and public sectors, as well as the social mission of the nonprofit. In other words, they have become increasingly hybrid in their operations.

For simplicity, we may think of nonprofit social housing organizations in the US as arising out of a social mission that focuses on 'people', 'places', or 'projects'. Regardless of their original orientation, they embrace the goal of providing decent, affordable housing to lower-income residents over the long term. Hybridity captures the modes of operation that organizations have pursued (sometimes out of necessity) to accomplish their mission-driven goals. (See Smith, 2010 for a discussion of the various types of nonprofit-based hybrid organizations in the US)

After providing an overview of the US social housing sector, this article addresses several interconnected concepts and relationships, which are presented in Figure 1. The final section of the paper offers observations aimed at enhancing our understanding and support for nonprofit social housing organizations.

The US Social Housing Sector

At various points in time, many of the activities of the current generation of nonprofit social housing organizations were carried out by either the public sector or the nonprofit sector. The first major affordable housing initiative of the US—the public housing program—was authorized by the US Congress in 1937 and, in the decades of the 1950s, 1960s and 1970s, over 1 million units were produced (Bratt, 1989), which represented about 80 per cent of the total number of public housing units ever built. In the early days of the program, ownership, management and financing were all handled by the public sector.

In contrast to the relatively pure public housing model, which was a creation of the state, even the earliest nonprofit housing organizations were hybrids. While the early tenement house developers were motivated to house the poor and upgrade their living environments, they relied on much more than private philanthropy. Lawrence Friedman, one of the best chroniclers of early US housing reform movements, noted that most types of nonprofit initiatives relied on private sector involvement. In fact, as of 1919 there were only two purely philanthropic housing ventures—model tenements that were expected to yield no profit, rather than a limited profit. By encouraging investment by 'hard-boiled businessmen', profit, albeit limited, was central to housing the poor. Thus, the prevailing view was that 'the free market should eventually be able to solve the housing problem, except for those people who by virtue of ignorance or inability to bargain could not effectively make use of the market' (Friedman, 1968, pp. 77, 78, 80).

Despite the assumptions and high hopes behind the model tenement movement, Friedman concluded that it had failed since the business of running houses for the poor had not attracted reputable businessmen (1968, p. 85). Yet, it is clear that even the very earliest nonprofit social housing organizations in the US had a hybrid mode of operation. Perhaps, then, it is not surprising that the current generation of groups has been embracing elements of the private for-profit sector. Still, it is timely to question what has prompted the recent movement to hybridity.

The US social housing sector comprises two broad types of units: those owned and managed by local public housing authorities and those by nonprofit social housing organizations. This article focuses on the latter. Although the units owned by the 3000 public housing authorities represent almost one-half of this sector, as public entities they traditionally show far fewer hybrid characteristics. Nevertheless, as discussed in the article by Nguyen *et al.* (2012), here, too, we are witnessing increasing hybridity.

Millions of housing units that are federally subsidized, but that are not permanently removed or protected from private housing market price increases, are excluded from the social housing category. Thus, the approximately 2 million units that are subsidized through the Housing Choice Voucher program (previously known as the Section 8 certificate/voucher program) are not considered part of the social housing stock, since the subsidy is linked to the household, not the unit, and there are no long-term guarantees about how long the subsidy will be available to a given household. Other units that have been federally subsidized, but that have been produced by private for-profit developers, are also not considered social housing, since they must remain affordable only for a limited period of time. For example, units produced by private for-profit developers through the Low Income Housing Tax Credit (LIHTC) program are not viewed as part of the social housing sector since affordability is required only for 30 years, up from the original 15 years when the statute went into effect. However, under certain circumstances, developers may opt out of the full 30-year obligation. (Depending on individual state laws, required restriction periods may be longer than 30 years.) In contrast, LIHTC units produced by nonprofits are viewed as part of the social housing sector. Although the federal affordability time limits are the same for nonprofits as for for-profit developers, the former are committed to maintaining affordability over the long term.

There is no single source of information concerning the number of nonprofit housing organizations and the number of units owned by these groups, or under long-term affordability restrictions in the case of homeownership units. However, on the basis of work by Stone (2006), as well as other work by this writer (Bratt, 2012), I estimate that there are about 5 million units in the US social housing sector (see Table 1). This equals about 5 per cent of the total US housing stock. By way of comparison, of the 20 European Union countries (out of 27 members), for which data are available for any period from 2000–2008, nearly two-thirds (13) have higher percentages of their housing stock in the social housing sector; in 9 member countries between 12 and 32 per cent of the housing stock is classified as social housing (Dol & Hafner, 2010, p. 67).

Table 1. US social housing sector: approximate number of units (2011)

	Number	Per cent
Public housing		
Federal (family and elderly)	1 160 000	23
Department of Defense (for military households)	400 000	8
Other programs (state and local)	700 000	14
Total public housing	2 260 000	45
Nonprofit housing		
Community development corporations	1 252 000*	25
Housing Partnership Network members	231 000	4.5
Community land trusts	5 000	0.1
Limited Equity Cooperatives	425 000	8.5
Federally subsidized housing for the elderly	200 000	4
Habitat for Humanity	30 000	0.6
Mutual housing and other nonprofit organizations	240 000	5
Total nonprofit housing	2 383 000	48
		(rounded)
Estimated additional production since data collected on which above figures are based	~ 357 000	7
Total	~ 5 000 000	100

*According to a report released by the National Alliance of Community Economic Development Associations, as of June 2010, the production by community development corporations (CDCs) and large nonprofit housing producers totaled 1 614 000 units. The report also points out that the survey information (on which the CDCs number in the above table are based) also counted the large housing producers. Therefore, there may be some double counting in the number of units listed for CDCs and for Housing Partnership Network members. This is one reason why the above figures should be viewed as approximations.
Additional Sources: National Congress for Community Economic Development, (2005; Stone, 2006; Sard & Fisher, 2008; Housing Partnership Network, 2011; Habitat for Humanity, 2011. Table originally prepared for (Bratt, 2012).

How nonprofit social housing organizations should be categorized has been a subject of considerable interest. For example, Crossan & Til (2009) have conceptualized a 'Social Economic Continuum' that encompasses a range of activities, as well as the types of social housing enterprises that carry out the tasks. Czischke *et al.* (2010) have proposed a three-part framework for classifying the various types of social enterprise housing organizations. Building off the work of Billis (2010); Brandsen *et al.* (2005); Crossan (2007); Crossan & Til (2009); and Gruis (2005, 2008), the authors discuss how the organizations can be described, their motivations and their behaviors. Another scheme, from a team of US researchers, used the number of units produced as the major distinguishing factor (Melendez & Servon, 2007).

In this article, nonprofit housing organizations of the US are divided into three major groups, based on their primary areas of concern—people, places and projects. The types are not mutually exclusive and many nonprofits have more than one focus; even though housing may be the central mission of most such groups, virtually all pursue other activities (Melendez & Servon, 2007; Vidal, 1992).

The largest group of nonprofit housing developers is composed of community development corporations (CDCs). CDCs are primarily focused on 'places'—they are typically committed to revitalizing an economically depressed and often physically deteriorated area and usually have community-based leadership. While housing development is the most prominent activity of these groups, they often focus on job creation and other economic development activities, as well as various types of social services, including programs for the elderly, non-English-speakers and children. The most recent survey providing an estimate of the number of CDCs found about 4600 such groups (National Congress for Community Economic Development, 2005).

Although the primary activity of most CDCs is housing development, they do not typically produce high volumes. As of 2005 more than one-half of CDCs (56 per cent) had produced less than 100 units over the life of the organization and only 20 per cent produced more than 25 units per year between 2001 and 2005 (National Congress for Community Economic Development, 2005). Indeed, even the name of these groups, 'community development corporations', captures the hybrid nature of their operations. Similarly, the trade association of the UK's nonprofit social housing organizations, the National Housing Federation, uses the tagline, 'In Business for Neighborhoods' (Mullins & Pawson, 2010, p. 206).

The second category of nonprofits includes groups whose primary mission is to produce a high volume of housing units, or 'projects' (for rental and for sale). Production is often not focused in a particular geographic area; units are dispersed across several neighborhoods, city, or region. Included in this group are the 98 members of the Housing Partnership Network (HPN, 2011); such groups typically have far larger portfolios than CDCs. A survey of 63 HPN members, revealed a median production of 1884 units; 11 groups had produced more than 5000 units (Mayer & Temkin, 2006).

Community land trusts (CLTs) are included in this category even though they do not develop a large number of units; their overriding mission is to provide affordable housing over the long term. CLTs are entities where the ownership of the land is held by the nonprofit that leases it for a small fee to owners of the buildings on the land. Although each CLT home is owned by the household leasing the land, the home cannot appreciate at the same rate as comparable private market homes. Instead, equity appreciation is based on improvements to the home and to a fixed inflation index. In this way, the CLT model enables low- and moderate-income families to build a modest amount of equity, while also preserving the affordability of these homes in perpetuity for future income-eligible households. As of late September 2011 there were 246 CLTs listed on the national CLT database (National Community Land Trust Network, 2011).

The final category of nonprofits includes diverse groups that are focused on 'people'— committed to meeting the housing needs of a specific subpopulation (e.g., the homeless, veterans, women who have left abusive relationships, or people with HIV-AIDS). In addition, some nonprofits have formed specifically to produce housing for their members, including unions and religious congregations. Also within this general category are tenant cooperatives and mutual housing associations, formed by residents of subsidized developments.

In exploring how nonprofit social housing organizations meet the challenges of the private market, while serving targeted resident groups and neighborhoods, I use the image of a Quadruple Bottom Line to capture the multiple objectives of organizations that develop and maintain social housing. This refers to the simultaneous need for the development to be financially and economically viable while also meeting social goals. Thus, the four components of the Quadruple Bottom Line are:

- financial viability of the development;
- social and economic needs of the residents living in the housing;
- a sensitivity to the way the housing fits into the larger fabric of the neighborhood and contributes to neighborhood viability; and
- the housing should be as environmentally sensitive and sustainable as possible, which involves minimizing the use of nonrenewable energy resources and striving to reduce transportation needs (Bratt, 2008b).

Whoever deserves 'bragging rights' for being first to use any of the terms pertaining to multiple bottom lines (see box), the concept has been little used in relation to nonprofit housing and the Quadruple Bottom Line is useful in exploring the challenges facing US nonprofit social housing organizations as their operations become increasingly hybrid.

Box: A Note on the Concept of Multiple Bottom Lines

In 1994, I co-authored a report in which we used the phrase 'double bottom line', the simultaneous need for housing nonprofits to be financially accountable while also meeting social goals (Bratt *et al.*, 1994). Subsequently, I expanded this notion to include two more elements— how the housing fits into the larger fabric of the neighborhood/contributes to neighborhood vitality (Bratt, 2008a) and, finally, how the nonprofit and the housing relate to environmental concerns (Bratt, 2008b), thus completing the Quadruple Bottom Line.

Until recently, I was oblivious about the extent to which others have used a multiple bottom line construct. Indeed, a Google search in early September 2011 yielded the following references to multiple bottom lines: Double Bottom Line = 38 800 000; Triple Bottom Line = 3 370 000; Quadruple Bottom Line = 645 000; Quintuple Bottom Line = 145 000.

Some, such as Dart (2004), cite Emerson & Twersky's (1996) articulation of the 'double bottom line' in relation to social service organizations developing and implementing for-profit enterprises. According to an article in *The Economist* (2009), 'triple bottom line' was coined in 1994 by John Ekington and the phrase has been used extensively in business; it has also been the subject of several books (Henriques & Richardson, 2004; Savitz & Weber, 2006; Willard, 2002). By the mid-2000s the multiple bottom line terminology had been embraced by community activist Carter (2006) who explained that the 'triple bottom line' involved positive returns for all concerned: developers, government, and the community. So much for originality.

Nonprofit Social Housing Organizations Add Activities to Fulfill Missions: A Preview of the Current Movement toward Hybridity

Nonprofit social housing organizations have a long record of responsiveness to the changing needs and opportunities in their communities. This section provides a background for understanding the moves of many contemporary nonprofits toward increasing hybridity in their operations. However, a key distinction between the adaptations of nonprofits operating in earlier periods and those in the current era is that the former seem to have been pulled toward change to better respond to their missions, whereas contemporary groups appear to be pushed to change in response to contextual constraints, particularly funding limitations, as discussed in the following section.

In the mid-1990s a colleague and I were engaged by the Ford Foundation to study how nonprofit housing organizations went about promoting 'self-sufficiency' for the residents in their developments (Bratt & Keyes, 1997). Beyond the central focus of the research,

[handwritten margin notes: governance driving change? or is change driving corporate governance? convergence to private sector model?]

we found that among the 21 organizations studied there had been considerable modifications in the types of programs that were being offered. Most often, organizations started with a clear focus on people, places, or projects. However, over time, organizations expanded their scope to explicitly address problems related to the other areas. In order for groups to fully respond to their missions and to address the 'self-sufficiency' aspirations of their clients, they developed a set of interrelated initiatives that went beyond their original range of activities.

For example, we found that organizations whose core mission pertained to people (e.g., the homeless) started their work with a focus on individual and household needs, such as helping to promote personal responsibility, developing skills in preparation for work and in providing services. However, they came to acknowledge that permanent housing in a stable neighborhood is a key component of a viable long-term solution. The following presents a synthesis of views gleaned from interviews with staff members at these organizations.

> In order for residents to become actively engaged in their own self-sufficiency, their most basic needs must be met . . . For homeless people, there is an immediate need to provide shelter . . . temporary housing is not enough . . . We need to be directly involved in providing housing . . . Services alone cannot provide the stability and security that comes with a permanent roof over one's head. Decent, safe and affordable housing is needed to allow the greatest benefit from the services being offered. (Bratt & Keyes, 1997, p. 56)

Other organizations had a longstanding focus on place (e.g., neighborhood revitalization), initiating various economic development and community-building activities, as well as housing development. They, too, realized that in order to continue to meet client needs, they had to assume various social service functions that had disappeared from the neighborhood; housing development was seen as part of an overall neighborhood improvement strategy.

> The logic of being place-focused inexorably leads organizations to the need to provide services and support for the people being housed. If residents require services, then organizations must either find providers in the neighborhood or provide the services themselves . . . most of the groups in this category took on housing as one of their first initiatives to improve the quality of place (including) . . . 'fixing up' the existing stock . . . (This) creates jobs in construction, rehabilitation, deleading and management. Housing is an excellent vehicle for learning concrete skills and enabling participants in the development process to achieve success. (Bratt & Keyes, 1997, pp. 57, 61)

Finally, there were the professional nonprofit housing developers—groups focusing on producing decent affordable housing 'projects'. They, too, came to acknowledge that their tenants might not be able to stay in their units without various types of services.

> A primary motivation for making the move from rental housing to services was to preserve the developments that the organizations had labored so hard to build. Disruptive tenants or people with serious problems can create a whole series of negative impacts for other residents. In addition, developments become financially

destabilized if rent arrears become widespread. It is often the resident managers who first become aware of tenants' problems, ranging from domestic violence to substance abuse, to poor parenting, to physical and mental illness.... even without pathology, tenants are prone to economic vicissitudes and job losses, with the possibility of negative implications for the economic survival of the development. (Bratt & Keyes, 1997, p. 70)

In a subsequent study of 12 members of the HPN, which includes many of the largest nonprofit housing organizations in the US, a similar pattern of organizational shifts in programming was identified within nearly all the groups (Bratt, 2006a; see also Glickman & Servon, 1998 for further discussion of changes that nonprofits make in their operations). The following are five short vignettes of organizations, all of which were primarily 'project' focused, but that significantly expanded their 'place' or 'people' programming.

- After *The Housing Partnership, Inc.* of Louisville, Kentucky, assumed ownership of a 220-unit distressed project in 36 buildings, it had the opportunity to rehabilitate 10 additional buildings in the area. This became part of a broad neighborhood-wide redevelopment scheme that, among the various efforts, has involved collaborations with the police with positive impacts on reducing crime.
- New York City's *Settlement Housing Fund* has played a significant role in neighborhood revitalization, beyond its housing projects. In particular, they helped develop a new supermarket in the lower east side of Manhattan and, in the Coney Island neighborhood of Brooklyn, they participated in the development of a 7800 sq. ft. health center for the elderly.
- In the wake of the 1980s collapse of Pittsburgh's steel industry, *ACTION-Housing Inc.* saw that foreclosures were a huge issue and brought all the key stakeholders together. With significant support from labor and churches, and with involvement from the steelworkers themselves, they started a program that assisted more than 2500 area residents to save their homes from foreclosure. This became a model for a statewide program developed by the Pennsylvania Legislature, which has helped more than 40 000 homeowners avoid foreclosure
- By the end of the 1980s, staff at the *Cleveland Housing Network* realized that strong resident programs were needed to supplement their development agenda. At that time, about 50 per cent of their clients were on welfare and there was a growing concern that many of these people would not succeed as homeowners. The organization focused on helping people get a job, as well as providing case management and services, including adult literacy, technology education and assistance in avoiding eviction.
- Up until 2001, *BRIDGE Housing Corporation* of San Francisco, California, was resistant to become involved with a services agenda, seeing itself primarily as a developer, owner and manager of affordable housing. However, in that year, *BRIDGE* conducted a survey of its residents; some questions asked how residents felt they could become more self-sustaining and better able to meet their personal goals. The results gave *BRIDGE* the confidence to start developing programs aimed at enhancing upward mobility for their target population.

These examples provide a sense of what pulled groups to add new activities and programs to their core missions and orientations. We observe that nonprofit social housing organizations are flexible, opportunistic and strategic. They tend to change their scope of services as they become aware of different needs among their residents and within their community. The exact ways in which each organization meets the challenges of a changed environment is, perhaps, less important than the reality that nonprofits have an extensive track record of adaptability and change. These examples also provide a context for understanding the contemporary moves toward hybridity, which involves nonprofit social housing organizations blending the roles typically associated with the public and market-based for-profit sectors.

Key Reason for Nonprofits Incorporating Private Market Operations: Changes in Public Funding for Affordable Housing

Just as the earlier generation of nonprofits worked hard to integrate the needs of people, places and projects, contemporary groups are creating their own unique approaches to navigating a vastly changed development environment. We are witnessing a great deal of fluidity in the nonprofit social housing sector, as organizations are adding new modes of operation, largely in response to public funding limitations both for housing and social services. Paralleling the US, Mullins & Pawson (2010, p. 201) have observed that the shift in resource dependency from public funding to private capital in the UK was 'perhaps the most significant determinant of hybridity'. In looking at housing associations in three countries (UK, US and Australia), Gilmour similarly stressed that as the public sectors in each country have undergone reform, sometimes referred to as New Public Management, *marketization* has been a key strategy, exemplified by the introduction of private finance for housing associations (2009, p. 22). And this, in turn, has contributed to the movement of housing associations becoming hybrids. Thus, as US nonprofit social housing organizations have become more enmeshed with the private market, the outcome has been greater hybridity.

In the late 1960s, the federal government started to experiment with alternative programs involving private for-profit developers. Under the public housing program, which was the dominant federal housing subsidy for low-income household during the 1950 and early 1960s, the federal government provided deep public subsidies for the construction (and later the management) of housing. The federal government was the funding source and a single type of entity—the local public housing authority—produced the housing. Departing from this model, the new public–private partnership programs offered federal subsidies to private for-profit and nonprofit developers to build housing for low- and moderate-income households. The programs involved below-market interest rate loans that, by 1968, were provided by private lenders, rather than the federal government, although the latter continued to subsidize the interest rate on the mortgage. This was the opening wedge of the public–private partnership era of affordable housing.

The below-market interest rate programs were followed by another short-lived public–private initiative involving a direct, more generous federal subsidy, known as the Section 8 New Construction/Substantial Rehabilitation program. Created in 1974, it also relied on both for-profit and nonprofit developers, with the former being more dominant. Nevertheless, many nonprofit social housing organizations benefited from this program, since the funding was relatively simple and generous. In addition, a number of other federal and state programs operating during the 1960s and 1970s provided targeted assistance to the fledgling nonprofit housing sector (Bratt, 1989).

With the demise of the Section 8 New Construction/Substantial Rehabilitation program in 1983, in large part due to its high cost, the era of deep federal subsidies for low-income housing was over. Indeed, Grønbjerg & Salamon (2002) identify the 1980s as a key point of demarcation, with the significant decline in federal funding for most nonprofit activities in the US. Smith (2010) also sees this as an important period in the move toward hybridity, with an increase in government contracts for services.

The early public–private partnership programs not only depended on direct public subsidies, but they also encouraged private investment by providing lucrative tax incentives. These 'back-end' indirect subsidies became a major engine for stimulating affordable housing development and they supplemented the assistance provided through Congressional appropriations, which were channeled through the US Department of Housing and Urban Development (Bratt, 1989; Schwartz, 2010).

In 1986, the LIHTC program was created, signaling a complete move away from direct federal appropriations for housing and an increased reliance on indirect subsidies provided through the income tax system. In order to make the units affordable to the target population, a number of funding sources must be combined. In addition, LIHTC is dependent on private financing that, in turn, has pushed nonprofits to ever greater degrees of hybridity. Although the nonprofit sector is allocated a certain percentage of tax credits, the fact that nonprofits must compete both with each other and with for-profit developers for tax credit allocations makes it essential that these groups have a high level of sophistication, with a professional understanding of market conditions. More than 25 years of experience with this program has largely been positive, with about 2.2 million units produced through 2009 (US Department of Housing and Urban Development, 2012).

With the ever-increasing dependence on the private market and for-profit developers, various problems have arisen, such as 'expiring use restrictions'. As developments built through the below-market interest rate programs matured, starting in the 1980s, the regulatory agreements with Department of Housing and Urban Development requiring that the housing be rented to low-income people came to an end; developers were free to rent to market rate to tenants or to sell the buildings. While nonprofits worked hard to continue renting at affordable levels, the for-profit developers often were eager to maximize their profits by allowing the rent restrictions on the developments to expire. This prompted federal legislation and a series of cumbersome and costly initiatives that attempted to safeguard the units for lower-income households (see Schwartz, 2010). However, they were only marginally successful and thousands of affordable units were lost.

A second example of problems arising from the public–private partnership in affordable housing surfaced in the aftermath of the mortgage crisis. With a decline in the economy, investors' needs to purchase tax credits to offset income declined. As a result, investors virtually disappeared and projects that were dependent on the capital that was to be raised by the sale of the credits became infeasible. While the LIHTC's heavy reliance on the private sector has been credited with its success, this experience is a reminder that the dependence of affordable housing development on the private market can be tenuous. Further, as discussed later, the LIHTC program also presents challenges for nonprofits.

Several intermediary organizations, which are hybrids themselves, have partially replaced federal funding for nonprofit social housing organizations. NeighborWorks America has received considerable appropriations from the US Congress, and the Local Initiatives Support Corporation and Enterprise Community Partners have been funded by private corporate and philanthropic donations. Playing a number of roles, including providing technical assistance, advancing loans and grants, and helping local groups to sell tax credits to private investors, the nonprofit intermediaries 'help CDCs access market resources that these organizations could not access on their own' (Cohen, 2008, p. 73).

In summary, the end of deep federal housing subsidies and the increased reliance on private financing has required nonprofits to embrace 'market discipline' and, as a result, they have become more hybrid in their operations.

Mediating the Private Market and the Quadruple Bottom Line

An overarching observation summarizes the essence of the conflicts faced by increasingly hybrid nonprofits that are simultaneously attempting to achieve all four elements of the Quadruple Bottom Line: *Nonprofits are committed to serving an array of social, economic and neighborhood goals that are not typically met through market-based strategies. But with less federal funding, they are forced to become more market-oriented and to rely on the very actors and institutions that have neglected the nonprofits' target constituents and communities.*

Nonprofits typically serve clients who have been hit hard by the loss of jobs, often due to deindustrialization, and/or who are unable to find jobs paying a livable wage, perhaps because of inadequate training or educational attainment, a downturn in the economy, or the goal of employers to keep wages as low as possible. The target areas in which nonprofits work also are likely to be facing a host of problems, including crime and vandalism, that could result in increased construction and management costs and difficulties locating suitable tenants even for reduced rent properties. In the face of these challenges, nonprofits attempt to operate high-quality developments, often with resident and community services. In contrast, for-profit developers are geared to maximizing profits for the benefit of owners and shareholders and the financial bottom line is the major concern.

The potential conflict faced by nonprofits struggling to meet the needs of their residents while, at the same time, carrying out tasks associated with protecting the financial and real assets of property ownership was identified nearly four decades ago by a small advocacy organization: 'the rights and needs of the tenants come last, only after all other parties involved in the development have been satisfied' (The Housing & Community Research Groups, 1973, pp. 39, 41).

Others have noted the challenges CDCs face in both serving community needs, while doing business with the financial institutions, and governmental agencies that support development. CDCs and other nonprofits in the current era are less likely to protest discriminatory bank practices, for example, while they are dependent on them for financing. If there is too much divergence between those providing funds and the community, the nonprofit may not be able to generate sufficient financial resources (Garn *et al.*, 1976, p. 131). Berndt (1977) continued this analysis by pointing out the problems inherent in the operations of CDCs. Nearly a decade later, Schuman (1986) observed that 'there is at least an element of irony in the attempt to solve housing problems by reinforcing the tax-shelter and investment mechanisms that maintain a housing system based on profit rather than need' (1986, pp. 469–470; see also Bratt, 2006b).

In recent years, Stoecker (1997) has, perhaps, done the most work exploring the built-in contradictions between the work of CDCs and their position within the US capitalist system. His argument posits that the fundamental conflict between capital and community introduces a virtually insurmountable obstacle to CDC success. He stated:

CDCs manage capital like capitalists, but do not invest it for a profit. They manage projects but within the constraints set by the funders. They try to be community oriented while their purse strings are held by outsiders. They are pressured by capital to produce exchange values in the form of capitalist business spaces and rental housing. They are pressured by communities to produce use value in the form of services, home ownership and green spaces. This is more than a 'double bottom line' (Bratt *et al.*, 1994). It is the internationalization of the capital-community contradiction and it leads to trouble. (1997, p. 6)

Implicit in this argument is that the hybridity of nonprofits thwarts their ability to meet the Quadruple Bottom Line. Despite the community orientation and the roots of many groups in local protest movements, Stoecker (1997) has further suggested that some CDCs have become disconnected from their constituents and have ceased being good advocates for community concerns. Thus, nonprofits' financial dependence on the private sector and the ensuing hybridity also is making them less responsive to the resident and neighborhood components of the Quadruple Bottom Line. More generally, Eikenberry & Kluver (2004) have questioned the extent to which the increasing marketization of nonprofits has compromised their focus on democracy and citizenship.

In contrast, other analysts have noted that nonprofits demonstrate impressive abilities at mediating the needs of capital and community. For example, CDC housing is viewed as a good basis for enabling tenants to exercise control over their environments and CDCs often attempt to respond to the personal challenges facing tenants (Leiterman & Stillman, 1993, pp. 30–31; Goetz & Sidney, 1994). And, rather than becoming a pawn of capital, CDCs can

> be very successful at altering the political balance of development politics in the city, at changing the calculus of capital investors, and at enhancing the social and psychological fabric of a neighborhood . . . [In their] relationships with government and private developers, [CDCs] have become more, not less, powerful, as outside forces have become co-dependent on the political and economic resources that CDCs can 'bring to the table'. (Robinson, 1996, pp. 1648, 1660)

Rubin (2000) offered this sweeping and very positive assessment of the role that community-based development organizations play in a capitalist economy. Implicit here is the view that any problems arising from hybridity are surmountable:

> [They] tie together a social mission with capitalist realities . . . They follow a pragmatic ideology of a humane capitalism for social change that community members, as well as politicians of both conservative and liberal leanings, can accept . . . [They] are about increasing social equity and demonstrating that a market economy still can have a heart. By reshaping the underlying symbols of a capitalist society to serve a social need, [they] end up renewing hope in neighborhoods of despair. (Rubin, 2000, pp. 273, 274)

Thus, while some analysts feel that it may be virtually impossible for nonprofits to fulfill the various components of the Quadruple Bottom Line, or that conflict and stress are likely, others feel that it is possible for groups to meet their multiple goals, as captured by the upbeat comment, noted earlier (see box on p. 7), by Majora Carter (see also Bratt 1997). Nevertheless, such results typically require a great deal of hard work, high levels of trust and collaboration, and, in some cases, greater resources.

The following explores further the relationship between the hybridity of nonprofits and their efforts to mediate the private market and to fulfill all components of the Quadruple Bottom Line; the four key points are taken from the schematic overview shown in Figure 1. The discussion draws on some of my prior empirical work, as well as analyses by Glickman & Servon (1998); Leiterman & Stillman (1993), Vidal (1992), and Walker *et al.* (1995). Other specific sources are cited in the text. This section does not explore the extent to which the three types of nonprofits—those with primary orientations on people, places, or projects—respond differentially to the challenges of meeting the various components of the Quadruple Bottom Line and the comparative strengths and weaknesses of each approach. While this could be an instructive area for future research, it is beyond the scope of this article.

As Nonprofits' Missions Broaden, the Potential Conflicts Inherent in Trying to Meet All Components of the 'Quadruple Bottom Line' Become Apparent

The more distinct areas of concern that a nonprofit embraces as part of its mission, the more it will be challenged to meet multiple bottom lines. For example, a project-focused nonprofit may run into conflicts when it attempts to incorporate an environmental agenda and produce developments consistent with smart growth principles (e.g., located in areas that are serviced by public transportation). However, since such sites are likely to be more costly to acquire, the higher costs will lead to a greater need for subsidies and the likelihood that more financing sources will be needed.

Similarly, as nonprofits increasingly embrace environmental concerns, the more they will have to do sophisticated cost–benefit analyses. Green building design is seen as a way for nonprofits to reduce long-term costs to the organization, in the case of rental developments, and to homeowners, in the case of for-sale housing. However, green design may be more costly upfront, thereby requiring additional funds to complete the project.

Another example demonstrates that the financial viability of a development could be compromised if there is too much pressure from a local community. Thus, a place-focused nonprofit may decide to launch a project even if it is undercapitalized to establish local credibility or to respond to strong desires by city government officials or other key stakeholders. Without sufficient capital built into the deal, financially viable management of the development, over the long term, may be compromised. The pressure to 'do something' to improve a distressed or vacant property can be acute, particularly if it has been a focus for illegal activities. However, the nonprofit's limited financial resources to tackle a particularly challenging property, even if moving forward appears to be the 'lesser of two evils', could mean that a project is undertaken that, perhaps, should not be (Bratt *et al.*, 1994; see also Rohe *et al.*, 2001).

There can also be conflicts between resident needs and project finances. Maintaining housing as affordable for lower-income households for as long as possible is virtually always a central goal of project-focused nonprofits. However, the social and economic needs of existing residents to pay affordable rent levels could conflict with the need for the nonprofit to raise rents as a result of increases in operating costs. If operating needs are short-changed, the financial viability of the project could become tenuous.

Concerning homeownership units, nonprofits have grappled with a key dilemma: if housing affordability for future owners is guaranteed, existing owners will not be able to take advantage of market appreciation and the resulting increases in equity. Many nonprofits have managed to resolve this dilemma by limiting the amount that equity can appreciate for a set period of time (often 5 years), after which owners are free to sell the property at market prices. Other housing nonprofits, notably those organized as CLTs, are able to provide long-term affordability, while still allowing individual homeowners to benefit from some modest asset appreciation.

In Attempting to Meet the 'Quadruple Bottom Line', Nonprofits are Likely to Become More Hybrid in their Operations

As nonprofits have confronted the desire to satisfy multiple bottom lines, they have had to become more savvy about affordable housing finance and about social service delivery systems. As discussed earlier, it is no longer possible to be involved with affordable housing development by securing a single deep subsidy from the federal government. And, as noted below, using the LIHTC program, for example, requires a high level of knowledge and sophistication about housing finance and the more the nonprofit masters these skills, the more hybrid it becomes.

Federal Funding Cutbacks put More Pressure on Nonprofits to Become More Hybrid in their Operations

Using the LIHTC program is a complex undertaking. In order to help nonprofits navigate the complexities of these deals, the large nonprofit intermediaries, mentioned earlier, provide assistance. But for nonprofits to develop housing that is affordable to lower-income households, they must develop expertise that allows them to participate in financing arrangements that rival (or surpass) the expertise needed to work in the private market.

A second push toward hybridity is that affordable housing is now seen not only as an end in itself, but also as a key component to assist low income households establish an economic foothold and move to a more secure family and work situation. Whether services are provided directly by the owner of the housing, or through networking with other agencies, nonprofits are often deeply involved with a broad range of resident-focused programs and, in fact, many nonprofits often see this as a key distinguishing characteristic between them and for-profit developers (Bratt, 2006a; Bratt, 2008b). Thus, some hybridity among nonprofits has involved offering an array of social services (either directly or by partnering or networking with others in the local community), following declines in funding for these programs and increasing difficulties faced by many social service providers to stay in business. For example, as this article was being completed, January 2012, Hull House, one of the US's earliest and most established social service agencies, located in Chicago, Illinois, announced that it was closing (*The New York Times, 2012*).

Increased Hybridity in Nonprofits' Operations Makes Fulfilling the 'Quadruple Bottom Line' More Challenging

Smith (2010) has pointed out that 'hybrid structures may be very helpful for nonprofit organizations as they strive to respond to a rapidly changing environment . . .' (227). At the same time, hybridity can create challenges for nonprofits as they aspire to meet all four elements of the Quadruple Bottom Line. For example, a nonprofit social housing organization that has adopted a social service orientation may find itself trying to siphon off a development's operating funds in order to cover its service programs However, this could jeopardize the financial viability of a development or even of the organization. This provides a further reminder that, while managing to meet all the components of the Quadruple Bottom Line is possible, it likely can only be achieved through a great deal of hard work, creativity, and cooperation among the nonprofit and key partners.

An examination of how some for-profit developers of affordable housing have altered their operations to become more socially oriented and to meet the non-financial aspects of the Quadruple Bottom Line also would be a valuable way of exploring hybridity. For example, the social mission and community-oriented activities of the private for-profit developer, McCormack Baron Salazar, appear to mirror those of nonprofits. Their website stresses its holistic approach to improving the quality of life for area residents by helping to build social institutions and sustainable, safe and efficient communities with jobs and economic opportunity for all residents, while protecting the environment and conserving resources (McCormack Baron Salazar, 2011).

Concluding Observations

As recently as 2002, the US Millennial Housing Commission conveyed the sense that the public sector, for-profit businesses and nonprofit organizations were all distinct: 'Effective delivery of affordable housing relies on [each sector doing] what each does best' (28). While hybridity was already well underway at that time, the blurring of these sectors has become more pronounced as the changes in public funding for affordable housing have resulted in nonprofits becoming increasingly hybrid in their operations. This paper has delineated various conflicts facing these organizations, as they attempt to fulfill the Quadruple Bottom Line—a task that has become more challenging as hybridity has increased.

In view of the possibility of significant cutbacks to a wide range of domestic programs in the US, including some mainstays such as Social Security and Medicare, the outlook for affordable housing, in general, and social housing, in particular, is not encouraging. As a result, nonprofit social housing organizations will be pressured into providing a variety of services in order to safeguard their developments and to meet resident and community needs. Whether the original focus of the group is around people, places, or projects, there will be substantial demands for nonprofits to embrace all three orientations. Helping organizations to make the needed changes as smoothly and efficiently as possible presents opportunities for support by the public sector, as well as from the large nonprofit intermediaries.

The public sector could better assist nonprofits to meet the demands of the Quadruple Bottom Line and thereby become more effective producers and managers of affordable housing. Although the importance of nonprofits in providing affordable housing opportunities to millions of US households is well established, there is still nothing approaching a 'nonprofit-centric' system of support for nonprofit organizations, which would involve a comprehensive array of funding for operations, pre-development costs, construction and permanent financing, and long-term subsidies, as well as technical assistance (Bratt, 1998). The piece-meal approach to supporting nonprofits notwithstanding, there have been many public (at the federal, state and local levels) as well as private philanthropic and other nonprofit-based efforts to sustain and enhance the work of these organizations.

State and local governments are in a particularly good position to assist nonprofits and many programs have been created to fill the federal void (Bratt, 1989; Goetz, 1993). In addition, as noted earlier, some members of the private for-profit development community are aspiring to 'do well by doing good'. As nonprofits and for-profits form partnerships, it would be desirable for both to acknowledge the goals, constraints and key motivations of each (Bratt, 2008b). In particular, understanding the multiple objectives of nonprofit social housing organizations, as represented by the Quadruple Bottom Line, provides a basis for developing the type of mutual understanding that is needed in productive nonprofit—for-profit partnerships.

As nonprofits continue to embrace new and more complex roles, increasing hybridity is a virtual certainty. At the same time, nonprofits will continue to try to meet all components of the Quadruple Bottom Line. Hopefully, this exploration will help the nonprofits, themselves, as well as the public and nonprofit agencies and funders that support their work, to acknowledge the competing objectives that organizations face and to develop strategies that serve to minimize the conflicts inherent in the Quadruple Bottom Line. Ultimately, the goal is to create a much more streamlined and efficient system for supporting nonprofits, often in collaboration with for-profits, to develop and maintain a robust stock of high quality housing that is affordable to lower-income households.

References

Berndt, H. E. (1977) *New Rulers in the Ghetto: The Community Development Corporation and Urban Poverty* (Westport, CT: Greenwood Press).

Billis, D. (2010) From welfare bureaucracies to welfare hybrids, in: D. Billis (Ed.) *Hybrid Organizations and the Third Sector: Challenges for Practice, Theory and Policy*, pp. 3–24 (London: Palgrave Macmillan).

Brandsen, T. W., van de Donk, W. & Putters, K. (2005) Griffins or chameleons? Hybridity as a permanent and inevitable characteristic of the third sector, *International Journal of Public Administration*, 28, pp. 749–765.

Bratt, R. G. (1989) *Rebuilding a Low-Income Housing Policy* (Philadelphia, PA: Temple University Press).

Bratt, R. G. (1997) CDCs: Contributions outweigh contradictions–A Reply to Randy Stoecker, *Journal of Urban Affairs*, 19(1), pp. 23–28.

Bratt, R. G. (1998) Nonprofit developers and managers: The evolution of their role in U.S. housing policy, in: C. T. Koebel (Ed.) *Shelter and Society: Theory, Research and Policy for Nonprofit Housing*, pp. 139–156 (Albany, NY: State University of New York Press).

Bratt, R. G. (2006a) *Housing Plus: Strategies, Challenges and Potential of Programs Provided by Members of the Housing Partnership Network* (Boston, MA: The Housing Partnership Network).

Bratt, R. G. (2006b) Community development corporations: Challenges in supporting a right to housing, in: R. G. Bratt, M. E. Stone & C. Hartman (Eds) *A Right to Housing: Foundation for a New Social Agenda*, pp. 340–359 (Philadelphia, PA: Temple University Press).

Bratt, R. G. (2008a) Viewing housing holistically: The resident-focused component of the housing-plus agenda, *Journal of the American Planning Association*, 74(1), pp. 100–110.

Bratt, R. G. (2008b) Nonprofit and for-profit developers of subsidized rental housing: Comparative attributes and collaborative opportunities, *Housing Policy Debate*, 19(2), pp. 323–365.

Bratt, R. G. (2012) Social housing in the U.S., in: S. J. Smith, M. Elsinga, L. Fox-O'Mahony, S. E. Ong & S. Wachter (Eds) *The International Encyclopedia of Housing and Home* (Elsevier).

Bratt, R. G. & Keyes, L. C. (1997) *New Perspectives on Self-Sufficiency: Strategies of Nonprofit Housing Organizations* (Medford, MA: Tufts University).

Bratt, R. G., Keyes, L. C., Schwartz, A. & Vidal, A. (1994) *Confronting the Management Challenge: Affordable Housing in the Nonprofit Sector* (New York: New School for Social Research).

Carter, M. (2006) Greening the ghetto: TED: Ideas worth sharing. Available at http://www.ted.com/talks/majora_carter_s_tale_of_urban_renewal.html (accessed 2 September 2011).

Cohen, R. (2008) Infrastructure in action: Bolstering nonprofit community developers, *The Nonprofit Quarterly*, 15(4), pp. 70–80.

Crossan, D. (2007) Towards a classification framework for not-for-profit organizations, Ph.D. dissertation, University of Ulster, Magee Campus: School of International Business.

Crossan, D. & Til, J. V. (2009) Towards a classification framework for not-for-profit organisations—The importance of measurement indicators, EMES Conferences Selected Papers Series, ECSP–B08-01.

Czischke, D., Gruis, V. & Mullins, D. (2010) *Conceptualizing Social Enterprise in Housing Organization*, (unpublished manuscript) (Istanbul, Turkey: European Network of Housing Research).

Dart, R. (2004) Being "business-like" in a nonprofit organization: A grounded and inductive typology, *Nonprofit and Voluntary Sector Quarterly*, 33(2), pp. 290–310.

Davis, J. E. (1994) *The Affordable City: Toward a Third Sector Housing Policy* (Philadelphia, PA: Temple University Press).

Defourny, J. & Nyssens, M. (Eds) (2008) *Social Enterprise in Europe: Recent Trends and Developments* (Liège, Belgium: European Research Network (EMES) Working Papers series. No. 08/01). Available at http://www.seeewiki.co.uk/~wiki/images/5/5a/SE_Strategy_for_success.pdf (accessed 10 September 2011).

Department of Trade and Industry (DTI) (2002) *Social Enterprise: A Strategy for Success* (Westminster, UK: Department of Trade and Industry).

Dol, K. & Haffner, M. (Eds) (2010) *Housing Statistics in the European Union*, The Hague: Ministry of the Interior and Kingdom Relations, OTB Research Institute for the Built Environment, Delft University of Technology.

The Economist (2009) The triple bottom line: It consists of three ps: Profit, people and planet, November 17. Available at http://www.economist.com/node/14301663 (accessed 10 September 2011).

Eikenberry, A. M. & Kluver, J. D. (2004) The marketization of the nonprofit sector: Civil society at risk? *Public Administration Review*, 64(2), pp. 132–140.

Emerson, J. & Twersky, F. (Eds) (1996) *New Social Entrepreneurs: The Success Lessons and Challenge of Non-Profit Enterprise Creation* (San Francisco, CA: The Roberts Foundation).

Friedman, L. M. (1968) *Government and Slum Housing: A Century of Frustration* (Chicago, IL: Rand McNally & Company).

Garn, H. A., Tevis, N. L. & Snead, C. E. (1976) *Evaluating Community Development Corporations—A Summary Report* (Washington, DC: The Urban Institute).

Gilmour, T. (2009) Network power: An international study of strengthening housing association capacity, Doctoral dissertation, Faculty of Architecture, Design and Planning, The University of Sydney, Sydney, Australia.

Glickman, N. J. & Servon, L. J. (1998) More than bricks and sticks: Five components of community development corporation capacity, *Housing Policy Debate*, 9(3), pp. 497–539.

Goetz, E. (1993) *Shelter Burden: Local Politics and Progressive Housing Policy* (Philadelphia, PA: Temple University Press).

Goetz, E. & Sidney, M. (1994) Revenge of the property owners: Community development and the politics of property, *Journal of Urban Affairs*, 16(4), pp. 319–334.

Grønbjerg, K. A. & Salamon, L. M. (2002) Devolution, marketization, and the changing shape of government-nonprofit relations, in: L. S. Salamon (Ed.) *The State of Nonprofit America*, pp. 447–470 (Washington, DC: Brookings Institute Press).

Gruis, V. (2005) Bedrijfsstijlen woningcoporaties; hulpmiddel bij het invullen van het maatschappelijk ondernemerschap [Organisational archetypes housing associations; support for shaping social enterprise], *Building Business*, 7, pp. 54–57.

Gruis, V. (2008) Organisational archetypes for Dutch housing associations, *Environment and Planning C: Government and Policy*, 26(6), pp. 1077–1092.

Habitat for Humanity (2011) Available at http://www.habitat.org/intl/na/218.aspx (accessed 10 March 2011).

Henriques, A. & Richardson, J. (2004) *The Triple Bottom Line, Does It All Add Up? Assessing the Sustainability of Business and CSR* (London: Earthscan).

Housing Partnership Network (2011) Available at http://www.housingpartnership.net/impact/accomplishments/ (accessed 8 November 2011).

The Housing and Community Research Groups (1973) *Community Housing Development Corporations: The Empty Promise* (Cambridge, MA: Urban Planning Aid, Inc.).

Koschinsky, J. (1998) Challenging the third sector housing approach: The impact of federal policies (1980–1996), *Journal of Urban Affairs*, 20(2), pp. 117–135.

Leiterman, M. & Stillman, J. (1993) *Building Community: A Report on Social Community Development Initiatives* (New York: Local Initiatives Support Corporation).

Mayer, N. & Temkin, K. (2006) *Housing Partnerships: The Work of Large-Scale Regional Non-Profits in Affordable Housing.* Draft Report (Washington, DC: The Urban Institute).

McCormack Baron Salazar (2011) Available at http://www.mccormackbaron.com/about/mission/truly-sustainable-communities-philosophy (accessed 7 September 2011).

Melendez, E. & Servon, L. (2007) Reassessing the role of housing in community-based urban development, *Housing Policy Debate*, 18(4), pp. 751–783.

Millennial Housing Commission (2002) *Meeting Our Nation's Housing Challenges* (Washington, DC: Millennial Housing Commission).

Mullins, D. & Pawson, H. (2010) Housing associations: Agents of policy or profits in disguise? in: D. Billis (Ed.) *Hybrid Organizations and the Third Sector: Challenges for Theory and Practice* (London: Palgrave Macmillan).

National Alliance of Community Economic Development Associations (2010) *Rising Above: Community Economic Development in a Changing Landscape* (Allliance, Washington, DC).

National Community Land Trust Network (2011) Available at http://www.cltnetwork.org/index.php?fuseaction=Main.MemberList (accessed 10 September 2011).

National Congress for Community Economic Development (2005) *Reaching New Heights: Trends and Achievements of Community-Based Development Organizations* (Washington, DC).

Nguyen, M. T., Rohe, W. M. & Cowan, S. M. (2012) Entrenched hybridity in public housing agencies in the USA, *Housing Studies*, 27(4), pp. 1–19 (to be updated).

The New York Times (2012) Chicago's Jane Addams Hull House to close, January 19. Available at http://www.nytimes.com/aponline/2012/01/19/us/AP-US-Hull-HouseClosing.html?scp=1&sq=Hull%20House&st=cse

Robinson, T. (1996) Inner-city innovator: The non-profit community development corporation, *Urban Studies*, 33(9), pp. 1647–1670.

Rohe, W. M., Quercia, R. G. & Levy, D. (2001) The performance of non-profit housing developments in the United States, *Housing Studies*, 16(5), pp. 595–618.

Rubin, H. J. (2000) *Renewing Hope within Neighborhoods of Despair: The Community-Based Development Model* (Albany, NY: State University of New York Press).

Sard, Barbara & Fischer, W. (2008) *Preserving Safe, High Quality Public Housing Should Be a Priority of Federal Housing Policy* (Washington, DC: Center on Budget and Policy Priorities).

Savitz, A. W. & Weber, K. (2006) *The Triple Bottom Line: How Today's Best-Run Companies Are Achieving Economic, Social and Environmental Success—and How You Can Too* (Chichester, UK: John Wiley and Sons).

Schuman, T. (1986) The agony and the equity: Self-help housing, in: R. G. Bratt, C. Hartman & A. Meyerson (Eds) *Critical Perspectives on Housing* (Philadelphia, PA: Temple University Press).

Schwartz, A. F. (2010) *Housing Policy in the United States*, 2nd ed. (New York: Routledge).

Smith, S. R. (2010) Hybridization and nonprofit organizations: The governance challenge, *Policy and Society*, 29, pp. 219–229.

Stoecker, R. (1997) The CDC model of urban redevelopment: A critique and alternative, *Journal of Urban Affairs*, 19(1), pp. 1–22.

Stone, M. E. (2006) Social ownership, in: R. G. Bratt, M. E. Stone & C. Hartman (Eds) *A Right to Housing: Foundation for a New Social Agenda* (Philadelphia, PA: Temple University Press).

Teasdale, S. (2010) What's in a name? The construction of social enterprise. Working Paper 46 (University of Birmingham, UK: Third Sector Research Centre, September).

U.S. Department of Housing and Urban Development (2012) Low-Income Housing Tax Credits. Available at http://www.huduser.org/portal/datasets/lihtc.html (accesssed 27 January 2012).

Vidal, A. (1992) *Rebuilding Communities: A National Study of Urban Community Development Corporations* (New York: Community Development Research Center, New School for Social Research).

Walker, C. et al. (1995) *Status and Prospects of the Nonprofit Housing Sector.* Report prepared by the Urban Institute for the U.S. Department of Housing and Urban Development, Contract number HC-5856.

Willard, B. (2002) *The Sustainability Advantage: Seven Business Case Benefits of a Triple Bottom Line* (Gabriola Island, British Columbia: New Society Publishers).

Entrenched Hybridity in Public Housing Agencies in the USA

MAI THI NGUYEN*, WILLIAM M. ROHE* & SPENCER MORRIS COWAN**

*Department of City and Regional Planning, University of North Carolina at Chapel Hill, Chapel Hill, NC, USA,
**Center for Urban and Regional Studies, University of North Carolina at Chapel Hill, Chapel Hill, NC, USA

ABSTRACT   *In this paper, we build on the extant literature on housing social enterprises and hybrid models of public housing delivery. We trace the evolution of US housing policy toward greater hybridity, focusing on three dimensions of hybridity. Drawing from a case study of the Charlotte Housing Authority in North Carolina, we showcase two housing programs, HOPE VI and Moving to Work, in order to highlight current innovations in the provision of housing for low-income populations and the entrenched hybridity that is evident. Using this information, we address two main questions: (1) how do local public housing agencies collaborate with the Federal government, private developers, and non-profit service providers to fund, construct, and manage affordable rental housing? and (2) what are the benefits and challenges of hybrid models of affordable rental housing delivery within the US context?*

Introduction

First built in 1930s, the development of public housing in the USA was a close collaboration between Federal and local governments. While the Federal government provided the construction funding, local government authorities owned, operated, and managed public housing with the mission of providing safe, decent, and sanitary living environments for selected families.[1] With the exception of contracting out design and construction duties to private architects and developers, the provision of public housing was purely a public endeavor.

Over time, the public housing program moved away from being a purely public to a hybrid endeavor—one that involves the public, private, and non-profit sectors and increasing inter- and intra-organizational complexity. This process of increasing hybridization was accelerated in the early 1990s due to two fundamental shifts. The first relates to increasing decentralization of authority over the provision of public housing,

53

resulting in local public housing agencies having greater flexibility in how they fund, construct, and manage public housing. As a result, hybrid approaches can be found in every stage of affordable housing delivery. Furthermore, decentralization has enabled local public housing agencies to craft context-specific policies and practices that may result in greater efficiency and higher-quality delivery of housing and housing-related support services. But it also poses some dilemmas, such as whether demolishing conventional public housing and replacing it with lower-density, mixed-income affordable housing developments serves fewer poor and vulnerable households.

A second shift relates to the reconceptualization of public housing from merely bricks and mortar, to a high-quality living environment with amenities and support services. Due to a dilapidated public housing stock and plaguing problems with persistent and intergenerational poverty, recent efforts by policy-makers have focused on breaking the cycle of poverty by providing service rich housing in scattered-site, lower-density, mixed-income, and mixed-use neighborhoods. Both of these shifts are often attributed to neoliberal views about the public provision of services (Drier & Atlas, 1996; Hackworth, 2005).

In this paper, we provide a brief history of the transformation of US housing programs toward greater intra-organizational, inter-organizational, and programmatic hybridity, with particular emphasis on the programs operated by local public housing agencies. We draw from a case study of the Charlotte Housing Authority (CHA) in North Carolina to provide concrete examples of these three dimensions of hybridity. We showcase two housing programs implemented by the CHA, HOPE VI and Moving to Work (MTW), in order to highlight current innovations in the provision of housing for low-income populations and the entrenched hybridity that is evident. Using this information, we address two main questions: (1) how do local public housing agencies collaborate with the Federal government, private developers and non-profit service providers to fund, construct, and manage affordable rental housing? and (2) what are the benefits and challenges of hybrid models of affordable rental housing delivery within the US context?

Social Enterprise, Hybridity, and the Delivery of Affordable Rental Housing

To provide a clearer understanding of our case study, we define 'social enterprise' and 'hybridity' as they relate to the provision of affordable rental housing[2] in the USA. In the broadest sense, social enterprises in the USA are organizations that engage in market-oriented economic activities while providing goods and services that serve a social mission (Czischke *et al.*, 2010; Defourny, 2009). Social enterprises apply principles of commercial markets to achieve greater efficiency, efficacy, and innovation in supplying public goods and social services (Kerlin, 2006, 2010).

In the USA, housing social enterprises can take on a variety of different organizational structures. Within the public sector, local housing providers have engaged in social entrepreneurial activities by creating for-profit and non-profit subsidiaries.[3] This allows them to draw from a larger pool of resources within the public, private, and non-profit sectors, take greater risks in their market-oriented activities, and have more flexibility. Private-sector entities also engage in supplying affordable housing, taking advantage of government incentives, such as the low-income housing tax credits (LIHTC)[4], which provide yearly tax write-offs for those who invest in approved affordable housing developments, which can make constructing, managing, and investing in affordable housing profitable. There are also a growing number of private non-profit housing organizations that fund, construct, and manage affordable housing. These organizations typically receive funding from the state, philanthropic organizations, or donations and employ business strategies to fill the gaps between the demand for affordable housing and what government entities can provide.

Given that social enterprises are present in the public, private, and non-profit sectors[5] and have a variety of ways in which they can structure their organization and management, housing social enterprises in the USA exemplify dimensions of 'hybridity'. Billis (2010) identifies three distinct sectors: public, private and third (or non-profit), which are distinguished by the following core elements: (1) ownership, (2) governance, (3) operational priorities, (4) human resources, and (5) other resources (such as financial). Hybrid zones emerge when there is some interaction between the three sectors. According to Billis (2010), there are nine possible hybrid zones (see p. 57 for a diagram).

In our case study, for example, the CHA falls within the public–private–third hybrid zone (or zone 2 in Billis, 2010), which has roots in the public sector, but currently operates under a hybrid model. Considering its hybrid governance structure, the CHA is governed by a Board of Commissioners that is appointed by local elected officials. One of the board members must be a tenant currently residing in a unit owned by the CHA. The CHA has created a wholly-owned, non-profit subsidiary which, in turn, has created a wholly owned for-profit subsidiary. Both the non-profit and for-profit subsidiaries are governed by CHA's Board of Commissioners, but they have different operational priorities. Both the CHA and its non-profit subsidiary pursue the social priorities specified in their corporate charters, such as providing housing for low-income households. The for-profit subsidiary, however, can seek to maximize profits. That different operational priority makes the for-profit subsidiary more attractive as a general partner to other for-profit investors in a development project than a non-profit or governmental entity.[6] Operating within these hybrid zones creates a 'blurring' of boundaries between ideal-typical organizations or sectors (Czischke et al., 2010).

Much of what has been discussed thus far in terms of hybridity can be considered institutional hybridity (Czischke et al., 2010), which relates to the hybrid structure within a single organization. For the purposes of our study, we refer to this dimension of hybridity as intra-organizational hybridity. We contrast this with another dimension of hybridity, which we refer to as inter-organizational hybridity, defined as the collaboration of hybrid organizations having roots in different sectors (public, private, and non-profit) to achieve a social mission. For example, inter-organizational hybridity in funding an affordable housing development would be when the hybrid local government agencies, private entities, and non-profit organization all collaborate and contribute funds to the development of an affordable housing development. We note that the key here is not simply collaboration across the various sectors, which has been studied extensively in fields such as public administration, but rather that mixed public, private, and non-profit organizations are working together to provide goods and services.

A third dimension of hybridity concerns the scope of services or goods provided, such as housing that includes social and support services. Others have referred to this as 'behavioral variables' (Crossan & Van Til, 2009). Housing social enterprises often achieve this form of hybridity by partnering with other organizations from different sectors. The partner organizations provide the additional services to their clients including: financial literacy courses, mentorship, workforce training, adult education, transportation, substance abuse rehabilitation, healthcare, and childcare. The result is that tenants living in affordable housing receive a host of goods, services, and amenities beyond the physical structure of housing. We refer to this dimension of hybridity as *programmatic* hybridity.

A final consideration is the changing nature of hybridity over time. Organizations may change their hybrid forms over time, decreasing or increasing their level of hybridity and, thereby also shifting across different zones of hybridity (Billis, 2010). Distinctions are made between *shallow* and *entrenched* hybridity in the third sector, where shallow hybridity starts to occur when third sector organizations initially start to adopt hybrid elements, such as hiring paid staff. Hybridity within these organizations become entrenched when the entire operation requires the hybrid model in order to function, as a result of becoming dependent on funding or services from a combination of sectors. Sometimes, organizations become entrenched 'organically' or through some evolutionary process and other times, they are 'enacted' or start out as entrenched hybrid organizations (Billis, 2010).

The Hybridization of Public Housing in the USA

The first large-scale public housing program in the USA was developed after the passage of the 1937 US Housing Act, also referred to as the Warner Steagall Act. The Act established the United States Housing Authority, a Federal agency that would be responsible for lending money and contracting with local public housing agencies to clear slums and build quality housing for the poor. The original Act made clear that the role of the Federal government was to enable state and local authorities to own and construct public housing by providing sufficient financing. The 1937 Housing Act stated,

> 'It is the policy of the United States to promote the general welfare of the Nation by employing its funds and credit, as provided in this Act, to assist the several States and their political subdivisions to remedy the unsafe and unsanitary housing conditions and the acute shortage of decent, safe, and sanitary dwellings for families of lower income and, consistent with the objectives of this Act, to vest in local public housing agencies the maximum amount of responsibility in the administration of their housing programs' Sec. 2. [42 U.S.C. 1437] (http://www.hud.gov).

Thus, conventional public housing was decidedly un-hybrid. It was initially a public endeavor involving Federal, state, and local governments. Because the Act required that state or local authorities be responsible for owning, building, and maintaining public housing, state and local governments established a multitude of local public housing agencies. The public housing program was also narrowly focused on producing subsidized housing units with limited attention to the social needs and aspirations of the residents.[7]

Twenty years after the passage of the Act, 840 local public housing agencies had been created by 1109 local jurisdictions in the country (Fisher, 1959).[8] The expansion of the public housing program continued for several more decades until President Richard Nixon placed a moratorium on new public housing construction in 1973 due to concerns over the expense of the program (Hackworth, 2005). This marked the beginning of an era of funding cuts and increasing decentralization of authority over public housing. This transition has been largely attributed to neoliberal views that promote greater individual autonomy, reliance on market mechanism, and less state intervention in the delivery of goods and services (Drier & Atlas, 1996; Hackworth, 2005).

Today, there are roughly 3300 local public housing agencies that have approximately 1.2 million housing units in their portfolio (http://www.hud.gov). Due to substantial cutbacks in Federal funds for housing and the adoption of neoliberal housing policies, many local public housing agencies have turned to social entrepreneurs to maintain their existing housing and, at least in some cases, add additional units. Varying forms of hybrid organizations have taken shape among local public housing agencies in the USA, yet, unlike Britain and some other European countries, there has been little discussion about how this hybridity emerged over time or its various dimensions.[9] There has been some work on hybridity within the non-profit sector in the USA, particularly relating to how the LIHTC has altered the governance, institutional organization, and mission of non-profit housing organizations, but little attention has been paid to housing authorities (e.g. see Smith, 2003, 2008, 2010).

The first widespread movement toward inter-organizational hybridity in public housing agencies in the USA came in 1974 with the introduction of the Section 8 Housing Allowance Program (later renamed the Housing Choice Voucher Program). Section 8 was intended to reduce both the cost of providing affordable housing and the concentration of poverty found in conventional public housing developments.[10] It provides LHAs with funds to assist low-income families in renting privately owned housing units. As such, it is an example of inter-organizational hybridity in the delivery of affordable rental housing: The public and private sectors work together to provide affordable housing. Under the program, private landlords are provided with the difference between 30 per cent of tenant incomes and unit contract rents. As with the conventional public housing program, LHAs administer the program under a set of detailed federal guidelines covering the condition of the private-sector units, the amount of the subsidy provided, the incomes and other characteristics of program participants, and many other topics.

Another important program that facilitated the hybridization of the public housing program was the 1989 passage of the LIHTC program. Although it is not managed by LHAs, they have used it extensively in the development of mixed-income developments. The LIHTC program encouraged private interests to make equity investments in affordable rental housing in return for fairly generous tax credits, which reduced the federal taxes owed by the investor dollar for dollar. In return, the housing units developed under this program have to be rented at a rate affordable to households making less than 60 per cent of the local median income for a minimum of 15 years. This program has become the main source of new affordable rental housing in the USA (Schwartz, 2010).

In the 1980s, selected LHAs began to move toward intra-organizational and programmatic hybridity. The introduction of the Urban Redevelopment Demonstration, later renamed HOPE VI, contributed to increasing hybridization of public housing delivery. In reference to the HOPE VI program, Katz (2009) notes: 'What started as an ambitious effort to revitalize the most distressed public housing has morphed into a full scale overhaul of the public housing program' (p. 27).

The HOPE VI program was designed to address the severely distressed public housing developments in the country: developments that were in extremely poor physical condition, in impoverished neighborhoods, and that had a variety of social problems including high crime. The program provides LHAs with funding to redevelop distressed public housing developments into attractive, mixed-income housing developments. The program's emphasis on providing mixed-income developments, coupled with the income restrictions as to who could reside in units funded by the conventional public housing program, required PHAs to partner with other private and non-profit housing providers. Rather than relying on the Federal government for all the funds needed to develop a project, the HOPE VI model required LHAs to attract private investment for the LIHTC and market-rate housing units in the new developments.

The additional demands created by mixed-income developments posed challenges for LHAs, as they were accustomed to administering Federal regulations, not engaging in entrepreneurial activities. As noted by Cisneros (2009),

> 'PHAs [Public Housing Authorities][11] did not have the organizational structure and trained personnel to execute real estate transactions of the complexity required under the HOPE VI model. For HOPE VI to be a success, the authorities would have to retain the requisite talent, revamp their accounting systems, adopt less centralized organizational models, and learn the techniques of site management' (p. 10).

In short, LHAs have had to develop skill sets similar to private real-estate developers.

The emphasis on creating mixed-income developments also meant the ownership of both the developments and the units built with public housing funding needed to be reconsidered. Traditionally, all public housing units were owned and operated by LHAs. The U.S. Department of Housing and Urban Development (HUD) rules and regulations effectively prevented mixed-income communities. In 1994, however, HUD's General Council opined that there was nothing in the public housing enabling legislation that required public housing to be owned by LHAs, as long as private entities followed program regulations (Baron, 2009, p. 37). This enabled private firms to own and manage mixed-income communities that contained public housing units, with the LHAs providing the oversight to insure that the private entities comply with public housing rules and regulations.

The HOPE VI program also spurred programmatic hybridity by stressing the importance of providing both relocated and returning residents with services designed to support upward socioeconomic mobility. Traditionally, LHAs focused on the provision of housing while largely ignoring the social needs and aspirations of the residents. Providing tenants with employment related services—such as job preparedness, job training, and day care—was a very small part of what they did, if they did it at all. The provision of those services was seen as the responsibility of other community agencies.[12] The HOPE VI program, however, required participating LHAs to develop and implement a Community Social Services plan and a portion of all HOPE VI funding is designated for that purpose. LHAs typically contract with local social service agencies to provide those services, thus, greatly expanding both inter-organizational and programmatic hybridity. As of 2011, 132 LHAs have participated in the HOPE VI program and have redeveloped 254 public housing developments with $6.1 billion in federal funding (U.S. Department of Housing and Urban Development, 2011a, 2011b, 2011c, 2011d).

The HOPE VI program has not been without its critics. Some have argued that the program has led to the displacement of 78 000 households, has disempowered residents, and has contributed to a shortage of affordable housing (Crowley, 2009). Others have criticized the program for being inefficient when compared with the additional number of residents that could have been housed if the funds had been transferred to the Section 8 Housing Voucher Program (Utt, 2009). Finally, through more stringent tenant screening processes, the most challenged and needy tenants often do not qualify for residency, which raises concerns that HOPE VI projects are reducing the number of housing opportunities for the hard-to-house.

A second federal program that has enabled and encouraged hybridity in public housing is the MTW demonstration. MTW was authorized by Congress in 1996 to address criticisms that publicly assisted housing programs breed dependency, undermine participant work ethic, and trap participants in areas with limited employment and educational opportunities. The program affords selected LHAs the flexibility to design and test innovative approaches to providing decent, safe, and sanitary affordable housing. It allows LHAs to 'block grant' the funding they receive from the federal government. They can combine their federal funding for public housing operating subsidies, Section 8 vouchers (also known as Housing Choice vouchers),[13] and capital budgets into a single, flexible account. Traditionally, those funds could only be used for their designated purposes. Participating housing authorities can also apply to HUD for exemptions from the federal rules that govern both the public housing and the Section 8 programs.

The LHAs approved to participate in the MTW program are expected to use this regulatory and financial flexibility to further three goals: (1) provide housing opportunities more efficiently; (2) increase the housing choices of low-income households; and/or (3) assist households in achieving self-sufficiency. The MTW program does, however, require participating LHAs to insure that at least 75 per cent of the families assisted are very low income, to assist substantially the same number of eligible families, and to provide housing for a comparable mix of family sizes as would have been served absent of the program. As of 2011, 35 LHAs have been approved to participate in the MTW program.

The MTW program has fostered all three dimensions of hybridity among the participating LHAs. First, in exempting them from many of the rules that have traditionally governed the public housing and the Section 8 programs, LHAs have much greater flexibility in determining how best to address local needs. In this sense, they can act more like private organizations, albeit they are still constrained by the need to mostly serve low-income households. The flexibility offered by the program allows LHAs to be entrepreneurial in identifying critical local needs and addressing them in innovative ways.

Second, the MTW program has fostered inter-organizational hybridity by allowing LHAs to partner with both for-profit and non-profit organizations in the provision of affordable housing. For example, MTW agencies can provide housing units or housing vouchers to other organizations, such as a mental health provider, to manage as service-enriched housing.

Third, the MTW program has fostered programmatic hybridity by establishing the goal of assisting residents achieve self-sufficiency. LHAs are urged to offer a wide range of employment-related services to their tenants. The flexible funding allowed under the program allows LHAs to shift funds from traditional line items like capital improvements to fund social service agencies.

As with the HOPE VI program, MTW has its critics. Some argue that it reduces the number of families receiving housing assistance, since most MTW authorities have used their financial flexibility to shift funding from their Section 8 Voucher programs to social services and administrative costs (Fischer, 2011). Others are concerned that allowing LHAs more flexibility in utilizing federal funds is a pretext for cutting the overall amount of funding provided to LHAs (Fisher & Sard, 2006). Critics also point out that MTW was conceived and sold as a demonstration program, but it has not been implemented in a way that allows for rigorous cross-site evaluation (Lubell & Barron, 2007).

With the expansion of the Section 8 and LIHTC programs and the development of the Hope VI and MTW programs, a paramount shift had occurred in how publicly assisted housing was delivered by the late 1990s. Hybridity was now integral, as a passage in the 1998 amendment (in the Quality Housing and Work Responsibility Act) to the 1937 Housing Act illustrates:

> "... our Nation should promote the goal of providing decent and affordable housing for all citizens through the efforts and encouragement of Federal, State, and local governments, and by the independent and collective actions of private citizens, organizations, and the private sector" (http://www.hud.gov).

Federal housing legislation now explicitly recognizes the view that public entities are not solely responsible or capable of providing sufficient levels of affordable housing.

The CHA: Hybridity Exemplified in Two Housing Programs

As of 2010, the city of Charlotte had a population of 731 000, making it the seventeenth largest city in the USA. Since the 1980s, Charlotte's population has more than doubled and continues to be one of the fastest growing cities in the country despite the current national economic recession. Driving Charlotte's growth and prosperity is the banking industry, which is second in size in the USA only to New York city. Due to rapid growth, the city has had to address housing problems that face households at different income levels, not just the very poor (University of North Carolina, Charlotte, Urban Institute, 2010). Faced with these housing challenges, the CHA's mission has been broadened from developing and operating conventional public housing to leading, developing and executing community-wide strategies that meet the broad range of housing needs for families who cannot otherwise obtain conventional housing. In addition, according to their mission statement, the CHA works 'to support and encourage families to move up and out of public housing and to integrate them into the economic mainstream' (www.cha-nc.org). This exemplifies the prioritization of autonomy and self-sufficiency as goals for assisted housing tenants.

In order to adhere to the mission, the CHA, like other LHAs in the USA, has become increasingly entrepreneurial, operating more like a private business than public agency. Like private-sector entrepreneurs, the CHA identifies opportunities, such as partnering with a non-profit organization that provides housing and social services to the homeless, and it gathers and allocates resources in pursuit of those opportunities (Drucker, 1985). It also innovates and takes more risks such as providing escrow accounts for residents and implementing work requirements. Thus, the CHA has taken on central qualities of many private organizations.

As true of other local housing authorities around the country, the CHA is a government corporation. In comparison to a public agency, government corporations have greater autonomy from political pressure, and arguably less accountability to the public since they do not directly report to public officials. The CHA Board of Commissioners consists of two members directly appointed by the Mayor of Charlotte and five members appointed by the Charlotte City Council, one of which must be a resident of publicly assisted housing. As mentioned earlier, the CHA also operates a non-profit and a for-profit subsidiary, both having different governance structures. Considering what Crossan & Van Til (2008) call 'descriptor variables', (e.g. baseline activities, legal structure, number of board and community members, and sources of funding) the CHA has become an entrenched hybrid organization. In other words, it could not continue its current operations and provide affordable rental housing as a pure public organization. Furthermore, the hybrid governance structure allows the CHA to prioritize different objectives, sometimes those that stand in conflict with one another, such as the trade-offs between autonomous governance and stakeholder involvement (Czischke, 2009; Czischke *et al.*, 2010). Our case study of two housing programs, HOPE VI and MTW, will show how the CHA has become an entrenched hybrid organization and offers examples of intra-organizational, inter-organizational, and programmatic hybridity.

The Park at Oaklawn: A Hope VI Redevelopment Project

Prior to the 1965 Housing Act, public housing in the USA could legally be racially segregated. When the Fairview Homes development opened in 1941, it was Charlotte's first public housing development for black residents. The development consisted of 410 family units in one- and two-story, barracks-style brick buildings and was located away from the city center. Over time, Fairview Homes deteriorated, and eventually the CHA determined that renovation of the obsolescent buildings would not be as cost-effective as demolishing them and building replacement units. Thus, in 1998 the CHA applied for and received a HOPE VI grant for $34 724 570 to demolish and replace Fairview Homes with a new development called the Park at Oaklawn.

The Park at Oaklawn project created mixed-income housing and diversified the clientele traditionally served by the CHA. The 410 family public housing units on the original site were replaced with 89 rental and 25 ownership units for family households and 83 units for seniors. Tenants in these units earn no more than 30 per cent of area median income, the target income level for conventional public housing. The redevelopment plan also included 89 rental and 46 ownership family housing units for households with incomes of up to 60 per cent of area median income, including 71 single-family and duplex units. The on-site redevelopment also included a 22 400 square foot community center as a place for residents to meet and access services. In addition to the on-site units, the Charlotte Mecklenburg Housing Partnership acquired seven off-site properties, adding 149 additional family and 192 senior public housing units, 344 family units and 26 senior units for households with incomes up to 60 per cent of area median income, and 304 market-rate units (see Table 1).

HOPE VI redevelopment projects, such as the Park at Oaklawn, are examples of what Billis (2010) might call 'enacted' intra-organizational hybridity. Compared with the older model of conventional public housing development, HOPE VI projects require a mixed-financing model. Therefore, the CHA played the role of financier and development partner in a complex, mixed-income development project, which included public housing, affordable housing, and market-rate units. This required that the CHA staff acquire different sets of skills and draw from the for-profit sector to create a new development model.

Although the CHA owned the redevelopment site, the development team exhibited the inter-organizational hybridity that is typical of HOPE VI projects. The Charlotte-Mecklenburg Housing Partnership, a 501(c)(3) non-profit corporation, was the main developer, not the CHA. Crosland, a for-profit corporation, developed, built, and currently manages the rental apartments and homes. Saussy Burbank, another for-profit

Table 1. Type of housing units, on- and off-site, Park at Oaklawn redevelopment plan

	Subsidized[a]	Affordable[b]	Market	Total
On-site units				
Family rental	89	89	0	178
Ownership	25	46	0	71
Seniors	83	0	0	83
Total on-site units	197	135	0	332
Off-site units				
Family	149	416	184	749
Senior	192	26	0	218
Total off-site units	341	442	184	967
Total family rental	238	505	184	927
Total ownership	25	46	0	71
Total senior	275	26	0	301
Grand total	538	577	184	1299

Notes: [a] Subsidized units are rented to households making equal to or less than 30 per cent of the area median income; [b] Affordable units are rented to households making equal to or less than 60 per cent of the area median income.
Source: Charlotte Housing Authority, (1998). The Park at Oaklawn Redevelopment Plan. Charlotte, North Carolina.

corporation, built and sold the site's single-family detached and duplex homes. The Mecklenburg County Parks and Recreation, a public agency, constructed a new community center.

This HOPE VI project is fairly unusual in that, through the acquisition of off-site properties, the number of units available to traditional public housing households was increased, rather than decreased. Most HOPE VI projects have led to an overall decrease in the number of public housing units available (Crowley, 2009). Considering both on- and off-site units, however, the total number of rental units for family households fell from 410 to 238, while the number of units reserved for seniors rose to 275. Not only did the redevelopment reduce the number of rental units for family households, it also reduced the number of units suitable for larger households. Almost 10 per cent of the units in Fairview Homes were either 4- or 5-bedroom units; less than 2 per cent of the units provided by the HOPE VI redevelopment project were that large (see Table 2). The result is that the replacement housing serves a larger, higher income, and more diverse group of tenants, which corresponds with recent changes in the CHA's mission of broadening the clientele base and providing additional community-wide amenities and services. However, this project reduces the number of very low-income and larger families served. It is unclear whether these reductions are due to concerns about the concentration of poverty in space or whether they are a product of this shift toward hybridity. CHA's reliance on federal funding and private developer's desire for future development contracts from the CHA (which are due to inter-organizational hybridity) create incentives for producing successful outcomes and having more very low-income and large families may threaten the success of a project and jeopardize future funding opportunities. Similarly, income mixing could be a response to addressing the negative consequences of concentrating poverty or evidence that the CHA is operating in a business-like fashion and making residents pay a higher percentage of unit costs, thereby reducing the CHA's overall costs.

Table 2. Family rental housing unit size, Fairview Homes and Park at Oaklawn, Charlotte, North Carolina

| | Number of bedrooms | | | | | |
	1	2	3	4	5	Total
Fairview Homes	72	205	93	28	12	410
Park at Oaklawn						
On-site	0	46	39	4	0	89
Off-site	30	84	35	0	0	149
Total	30	130	74	4	0	238

Source: Charlotte Housing Authority, (1998). HOPE VI Application for Fairview Homes. Charlotte, North Carolina.

The on-site rental units are managed by Crosland, the private company that built the apartments. The off-site rental units are managed by several different private-sector property management companies. For example, S. L. Nusbaum & Company manages two of the larger off-site developments, each with 20 public housing rental units, 80 affordable rental units, and 92 market-rate rental units. Those complexes offer amenities including swimming pools, fitness centers, club houses, and computer/business centers, which appeal to private market tenants. The use of private-sector managers reflects the inter-organizational hybridity necessary to entice market-rate tenants to live in mixed-income developments. The CHA has experience managing purely public housing developments, but it does not have experience managing market-rate units.

The services provided at the on-site community center exemplify intra-organizational hybridity. The center is where local non-profit organizations currently provide services to residents. For example, Bethlehem Center, an affiliate of the United Methodist Church, offers day care and the Head Start[14] program for young children, while the Anita Stroud Foundation, another non-profit, runs an after school program. By partnering with these two non-profits, CHA is able to expand the scope of services it can provide for its residents.

Funding for the Park at Oaklawn project came from a variety of public and private sources, illustrating the inter-organizational hybridity required by HOPE VI developments. The CHA contributed funds from the HOPE VI grant it received from the US Department of HUD and it issued tax-exempt bonds that were sold to private investors. CMHP contributed Section 202[15] funding from HUD for the senior housing. Wachovia Bank and other for-profit investors bought the tax credits and the tax-exempt bonds. The City of Charlotte provided infrastructure funding, while the Mecklenburg County Parks and Recreation Department contributed additional funding for the new community center. The original $34.7 million HOPE VI grant leveraged an additional $95.7 million, over half of which was private-sector investment (see Table 3).

The use of tax credits requires a private-sector entity to own the properties, and so the legal titles are held by for-profit corporations set up by non-profit partners, such as the Charlotte-Mecklenburg Housing Partnership. The tax-exempt bonds that the CHA issued were on behalf of the for-profit corporations that own the properties. While the legal structure and degree of control that the non-profit retains over the for-profit corporation may vary, in general, they will be sufficiently independent to insulate the non-profit from liability. In practice, the ownership structure is a hybrid mix. The owner is a for-profit

Table 3. Funding sources for the Park at Oaklawn HOPE VI redevelopment project

	HOPE VI ($)	Tax credits ($)	City ($)	Section 202 ($)	Other HUD/AHP ($)	Bonds or mortgages ($)	Total ($)
On-Site units							
Rental	6 195 159	8 077 617	577 200	0	0	0	14 849 976
Ownership	1 870 144	0	363 028	0	1 165 000	6 264 322	9 662 494
Senior	229 918	0	122 714	7 100 000	0	0	7 452 632
Total on-site	8 295 221	8 077 617	1 062 942	7 100 000	1 165 000	6 264 322	31 965 102
Off-site	14 603 003	28 124 345	11 411 639	0	4 452 818	28 075 000	86 666 805
Total	22 898 224[a]	36 201 962	12 474 581	7 100 000	5 617 818	34 339 322	118 631 907

Note: [a] The total Hope VI grant received was 34.7 million but the amount shown here does not reflect funds spent on tenant relocation, demolition, site infrastructure, client services, or public housing authority administration.
Source: Charlotte Housing Authority. (1998). The Park at Oaklawn Redevelopment Plan. Charlotte, North Carolina.

corporation that was established and is controlled by the non-profit organization. The for-profit corporation exists to help the non-profit organization serve its mission.

The CHA's MTW Program

In 2007, the CHA received HUD approval to participate in the MTW program as described earlier.[16] CHA's MTW program includes four major components: (1) rent reforms that apply to all households; (2) work requirements that apply to all able-bodied, non-elderly household heads; (3) support services to help households find and hold jobs; and (4) additional housing units in mixed-income developments, including more units for people with disabilities and special needs.

The CHA's work requirement is designed to promote economic independence by requiring all able-bodied, non-elderly residents to work an average of 30 h per week within 2 years of beginning the MTW program. Recognizing that many of its residents might have barriers that prevent them from being hired, the CHA developed programs that relied on greater inter- and intra-organizational hybridity. The CHA partnered with outside entities for resident services and it changed employee job descriptions and provided training to assist them in adapting to the new model.

The CHA, for example, partnered with the University of North Carolina at Charlotte to conduct an assessment of barriers to employment for every resident and it used the results of that assessment to target services—such as GED classes,[17] job and life skills training, and referrals to workforce development programs—to the residents that needed them to address employment barriers. Case managers have been assigned to all residents subject to the work requirement. Some of these case managers work directly for the CHA while others work for non-profit or for-profit agencies. Case managers assist residents in establishing goals and make referrals to local social service providers.[18] The decision to contract out service required CHA to change its own Client Services Division to focus more on contract management and administration, rather than on the direct provision of services.

One of the CHA's major MTW initiatives is to expand the number of supportive housing units it provides. To do this, rather than assuming its traditional role as landlord, the CHA is assuming the role of developer and financier and is partnering with outside organizations for services. For example, CHA is providing gap financing to St. Peter's Homes, a non-profit serving the homeless, for the construction of McCreesh Place, a 64-unit single room occupancy development, which will provide supportive housing for formerly homeless men with disabilities. St. Peter's Homes is developing the property and is part of a network of referral agencies and service providers who work with the homeless. In addition to the gap financing, CHA is also providing rent subsidies for residents at McCreesh Place.

For another project, the CHA is using its non-profit subsidiary, Horizon Development Properties, to acquire and redevelop the Hampton Creste development as a 213-unit mixed-income project, including 60 units of public housing, 50–55 units of low-income housing, and market-rate units. The public housing units will be targeted for the homeless, and the Salvation Army and other local service providers will offer supportive services to the residents of those units. The CHA's contribution is being supplemented by additional funding from the Critical Need Response Fund Task Force, a partnership among the Fund for the Carolinas, the United Way of Central Carolinas, Mecklenburg Ministries, the Leon Levine Foundation, and the Charlotte Chamber of Commerce.

As might be expected, Charlotte's Moving Forward program has its critics. Of particular concern is the introduction of the work requirement during a period of unusually high unemployment in the city. Critics worry that the public housing residents, even with additional job readiness and training programs, will not be able to find employment and, therefore, will lose their housing subsidy (Charlotte Observer, 1/5/09).

The MTW program is designed to both encourage and enable LHAs to assist residents in improving their economic and social circumstances. The participating LHAs are expected to expand their business models from a narrow focus on being landlords and property managers to a broader focus on assisting residents to increase their incomes by providing support services, such as job training. At the same time, the LHAs are expected to diversify their portfolios by constructing and/or acquiring units in more affluent and opportunity-rich neighborhoods, thereby deconcentrating poverty. To make those changes, the LHAs are partnering with outside entities in the public, non-profit, and private sectors that offer the expanded scope of services and housing options required by MTW.

The Federal government established both HOPE VI and MTW to address obvious problems that had arisen with the older business model for providing public housing. That model used federal funding to build units, and had LHAs own and operate the housing, without additional services to assist residents. While the model may have been adequate to meet the needs of working families, the original target population, it proved inadequate for the poorer, minority population that came to inhabit public housing after World War II. HOPE VI and MTW encouraged inter-organizational hybridity to form partnerships, first to leverage private-sector financing and then to provide the full array of services needed by public housing residents. Those two programs also encouraged LHAs to become more internally and programmatically hybrid, doing more for their residents and becoming more business like in their relationships with their partners. Staff at LHAs learned to structure complex financial deals to develop new units, and to assemble collaborative groups of providers to offer holistic service packages.

Benefits and Challenges of Hybridity

Since the Housing Act of 1937, the delivery of public housing in the USA has evolved from a pure public to a hybrid model, involving more complex organizational structures, inter-organizational collaborations and, tenant support services targeted to improve overall well-being, not just housing conditions. This evolution toward hybridity has encouraged Local Housing Authorities to harness the power of markets and employ business techniques, while also enlisting the services of non-profits, volunteers, and faith-based organizations to provide high quality affordable rental housing and tenant services. While there are many benefits of hybridity, challenges also arise with this new model of affordable rental housing delivery.

One benefit of this shift toward hybridity is the ability to leverage funding for costly housing projects. The CHA was able to leverage an additional $95.7 million—half coming from the private sector—with a $34.7 million Hope VI grant from the Federal government to redevelop the Park at Oaklawn. Involving private and non-profit sector organizations has led to substantial benefits in terms of both funding and expertise. For example, City Dive, a faith-based non-profit organization donated $2.3 million in volunteer mentor services to residents of a HOPE VI redevelopment project in Charlotte. The challenge for providers of affordable rental housing is to develop the necessary skills and expertise among staff members in order to apply for, coordinate, and leverage a substantial amount of funding from a variety of sources. After garnering the necessary funding, managing the funds appropriately and effectively requires experience. Ethical dilemmas may also arise in how to allocate money between various programs due to increased flexibility. Misuse of MTW funds by the Philadelphia Housing Authority, for instance, emphasizes the need for more oversight and monitoring by the Federal government and greater transparency and accountability by Local Housing Authorities as the process of affordable housing delivery becomes more decentralized and involves actors from a variety of different sectors (Office of Inspector General, 2004).

The privatization of the construction and management of affordable rental housing has certain advantages, but it may also pose challenges to the mission of providing housing to the poorest population. Public housing agencies often contract the design and development phase of an affordable rental housing project to private development companies that have experience with both market-rate and affordable housing development. Therefore, they are knowledgeable about working in the private as well as public housing market and apply the same architectural and design standards used in the private-market developments in their public housing developments. These experiences and skills are particularly beneficial in designing and developing mixed-income and mixed-use affordable housing projects, which are much more diverse in architectural style than historic, modernist public housing structures (Schwartz, 2010). LHAs can also contract with a number of private development companies to develop different types of housing (e.g. single-family detached, multi-family attached, or senior) within the same project in order to benefit from specialized development expertise.

Private property management firms have shown success in operating and maintaining more recently developed affordable rental housing by using the same standards in their affordable rental housing developments as their private market projects. Maintaining their reputation in the private real estate market and securing additional contracts with public housing agencies are the two reasons that motivate private management firms to adequately maintain their properties. Private management firms often have more restrictive criteria for tenant selection and have lower tolerance levels in regard to rule breaking that leads to evictions than management departments within LHAs. For example, Crosland, a private development company that manages a number of properties for the CHA, conducts criminal and background checks for adult residents and parents of minors must sign an affidavit that all their children do not have criminal records as part of the screening process.

The more restrictive screening and lower tolerance levels, however raise concern that the most troubled and needy residents are not being served under hybird rental housing programs, such as Hope VI and MTW. These programs may be 'creaming the crop' in order for both private developers and public housing agencies to report successful outcomes. An additional concern with private development companies having primary responsibility for design, development, and management is whether they are responsive to the needs of tenants who will live in the housing developments, the public at-large, or public officials and whether they adhere to the mission of the public housing agency.

What has been perhaps one of the biggest transitions from conventional public housing to modern affordable housing programs, such as HOPE VI and MTW, is the quality and amount of social services provided to residents to improve their overall quality of life. Our case study of Charlotte indicates that case managers, social service providers, and volunteers work closely with residents during the relocation process as they move out of conventional public housing while the new development is being built, during the construction, and afterwards, when some of the tenants return to the newly built structures. We still lack longitudinal data to know if residents of conventional public housing are able to achieve self-sufficiency due to the services they receive once they return to newly constructed HOPE VI projects or whether there are selection biases in the types of tenants that return. But what is evident is that there has been a fundamental shift in viewing housing as a means to provide both roots and wings—roots that allow affordable housing residents to settle comfortably into a home and neighborhood and wings that enable these residents to achieve socioeconomic mobility.

Discussion

Driven in large part by the neoliberal views shaping US public policy and cuts in Federal funding for public housing, hybridity in the delivery of affordable rental housing in the USA has become entrenched, both in federal housing programs and in many local public housing agencies. Our case study of the CHA illustrates how intra-organizational hybridity has allowed for more operational flexibility and funding resources due to different legal status designations of CHA subsidiaries. Inter-organizational hybridity has led to a more refined division of labor across organizations from the private, public, and non-profit sectors that allows for greater specialization and presumably efficiency in the delivery of affordable rental housing. Programmatic hybridity has allowed the CHA to contract with other organizations to provide a broad range of services designed to encourage and assist residents to strive for self-sufficiency.

The additional flexibility allowed to LHAs, under the newer hybrid programs, such as HOPE VI and MTW, has important potential benefits, including allowing creative solutions to housing lower-income households, permitting LHAs to customize their programs based on local conditions and priorities, and enabling a range of partnerships that have the potential to broaden the funding and organizational capacity needed to create new and better quality affordable housing. But this flexibility also comes with considerable risks, including shifting the focus away from the neediest families to more well-off ones as a means of ensuring financial solvency as HUD reduces funding for public housing programs (Devine et al., 1999; Fischer, 2011). The new flexibility also increases the likelihood of the inappropriate use of funds, as was recently seen in the Philadelphia Housing Authority.[19] To guard against those risks, clear program goals and performance standards are needed, as is more careful program monitoring by federal officials. Moreover, as hybrid models of affordable housing delivery are increasingly adopted by LHAs around the country, carefully designed evaluations are needed to assess the full range of program impacts.

Notes

[1] The original public housing program was designed to assist the 'submerged middle class', that is, two parent families where the breadwinner was out of work due to the depression. Single-parent families were not served by the program.

[2] We use the term affordable rental housing to refer to housing occupied by households who pay an affordable rate based on their income level. The US Department of Housing and Urban Development considers a unit affordable if a household pays no more than 30 percent of their income on rent. We do

not use the term public housing because of the shift in the supply and management of affordable rental housing by non-public entities. We also do not use the term subsidized housing because there are subsidies or grants given to homeowners as well as renters. Therefore, affordable rental housing could be public or subsidized housing but this is not always the case.

[3] It should be noted that there are other types of local public agencies that supply affordable housing. There are housing offices, divisions, or agencies that are housed within local governments. These entities are purely public agencies, therefore not social enterprises.

[4] Started in 1986, the Low-Income Housing Tax Credit Program is an indirect Federal subsidy to developers to assist in the financing of affordable rental housing development for low-income households. Developers who receive the tax credits can sell them to investors to raise capital (or equity) for their projects, which reduces the debt that the developer would have to borrow, and makes the investment less risky. Over a period of 10 years, investors receive a dollar-for-dollar credit against their Federal tax liability each year. For more details about the program, see http://www.hud.org.

[5] In other countries around the world, the non-profit sector is typically referred to as the third sector. In the USA, the term non-profit is more commonly used and, therefore we have chosen to use this term.

[6] Having a separate corporation as the general partner may also protect the parent corporation from liability. For example, if a development with the for-profit subsidiary as general partner failed and filed for bankruptcy, creditors would be repaid from the general partner's assets and the limited partners' assets to the extent they were invested in the development. Creditors should not be able to reach CHA's other assets, those not invested in the failed development, to satisfy the development's obligations.

[7] In limited instances, other organizations provided on-site services, such as health and recreation activities (Fuerst, 2005).

[8] Some local public housing agencies serve two or more jurisdictions.

[9] There has been more attention recently to hybrid models of housing delivery in European countries, particularly England (Buckingham, 2010; Pawson & Mullins, 2010; Teasdale, 2010) and the Netherlands (Mullins, 2011). There have also been comparative studies that examine historical and contextual differences in shaping hybrid housing organizations (see Czischke, 2009; Kerlin, 2006, 2009, 2010; Mullins & Pawson, 2010).

[10] An earlier example of hybridity in federal housing programs was the Section 236 rental assistance program under which HUD provided supply-side subsidies to the developers of affordable rental units in return for an agreement that the units developed be affordable to low-income households for a specified period of time, such as 15 or 20 years. We do not discuss that program in this paper since LHAs were not involved in administering that program.

[11] Public Housing Authority is often used interchangeably with Local Housing Authority.

[12] Before the HOPE VI program, HUD did provide limited funding for the Family Self-Sufficiency program but this program involved a small fraction of all public housing households.

[13] The Housing Choice Voucher Program provides households with vouchers to rent housing in the private market provided that the unit meets quality standards and does not exceed the maximum allowable rents set by the local housing authority. The program was designed to provide a wider range of neighborhood choices to low-income households in order to promote poverty deconcentration. To find out more about the program, see http://www.hud.gov.

[14] Head Start is an early childhood enrichment program serving disadvantaged children from birth to 3 years old.

[15] Section 202 funding can be used by non-profit organizations to construct, rehabilitate or acquire units to serve as supportive housing for the elderly with incomes below 50 per cent of area median income. The funds can also be used for rent subsidies for the units.

[16] While the nation-wide program is called 'Moving to Work', the CHA renamed its program, 'Moving Forward'. We will call the program 'Moving to Work' to be consistent with programs in other jurisdictions nation-wide.

[17] GED courses provide education on how to pass a GED exam, which would enable participants to obtain a degree equivalent to a high school diploma.

[18] The service providers include Genesis Project 1, Lutheran Family Services, Melange Health Services, Symmetry Behavioral Health Solutions, and Children's Home Society-Youth Homes Division.

[19] The Philadelphia Housing Authority was selected to participate in the Move to Work program and over several years used its financial flexibility to move $300 million out of its Section 8 program and used it for employee gifts, social events and excessive and unnecessary improvements to both administrative

buildings and housing developments. The authority has been placed under administrative receivership and there are ongoing investigations (Office of Inspector General, U.S. Department of Housing and Urban Development, 2010).

References

Baron, R. (2009) The evolution of HOPE VI as a development program, in: H. Cisneros & L. Engdahl (Eds) *From Despair to Hope: Hope VI and the New Promise of Public housing in American's Cities*, pp. 31–47 (Washington DC: Brookings Institution Press).

Billis, D. (2010) *Hybrid Organizations and the Third Sector* (Basingstoke: Palgrave).

Buckingham, H. (2010) Hybridity, diversity and the division of labor in the third sector: What can we learn from homelessness organizations in the U.K.? Third Sector Research Center Working Paper 50. Available at http://www.tsrc.ac.uk/

Charlotte Housing Authority (1998) Charlotte Housing Authority HOPE VI Application for Fairview Homes. Charlotte, North Carolina.

Charlotte Housing Authority (2008) The Park at Oaklawn Redevelopment Plan. Charlotte, North Carolina.

Charlotte Housing Authority (2011) Charlotte Housing Authority Mission Statement. Available at http://www.cha-nc.org

Charlotte Observer (2009) What tenants need is to help to get, keep jobs. Editorial page, January 5.

Cisneros, H. (2009) A new moment for people and cities, in: H. Cisneros & L. Engdahl (Eds) *From Despair to Hope: Hope VI and the New Promise of Public Housing in American's Cities*, pp. 3–13 (Washington DC: Brookings Institution Press).

Crossan, D. & Van Til, J. (2009) Towards a classification framework for not-for-profit organizations—The importance of measurement indicators, EMES-ISTR European Conference, Barcelona, Spain.

Crowley, S. (2009) HOPE VI: What went wrong, in: H. Cisneros & L. Engdahl (Eds) *From Despair to Hope: Hope VI and the New Promise of Public Housing in American's Cities*, pp. 229–248 (Washington DC: Brookings Institution Press).

Czischke, D. (2009) Managing social rental housing in the EU: A comparative study, *European Journal of Housing Policy*, 9(2), pp. 121–151.

Czischke, D., Gruis, V. & Mullins, D. (2010) Conceptualizing social enterprise in housing, European Network of Housing Research Conference, Istanbul, Turkey July 4–7.

Defourny, J. (2009) Concepts and realities in social enterprise: A European perspective, *Collegium*, No. 38 (Spring): 73–79.

Devine, D., Rubin, L. & Gray, R. (1999) *The Uses of Discretionary Authority in the Public Housing Program: A Baseline Inventory of Issues, Policy and Practice* (Washington DC: U.S. Department of Housing and Urban Development, Office of Policy Development and Research).

Drier, P. & Atlas, J. (1996) US housing policy at the crossroads: Rebuilding the housing constituency, *Journal of Urban Affairs*, 18(4), pp. 341–370.

Drucker, P. (1985) *Innovational Entrepreneurship: Practice and Principles* (New York: Harper).

Fischer, W. (2011) *Expansion of HUD's 'Moving-To-Work' Demonstration is not Justified* (Washington DC: Center on Budget and Policy Priorities). Available at http://www.cbpp.org/

Fisher, R. (1959) *20 Years of Public Housing: Economic ASPECts of the Federal Program* (New York: Harper and Brothers Publishers).

Fisher, W. & Sard, B. (2006) *Inspector General Reports on HUD'S Moving to Work Demonstration Raised Serious Questions* (Washington DC: Center on Budget and Policy Priorities). Available at http://www.cbpp.org/

Fuerst, J. S. (2005) *When Public Housing was Paradise: Building Community in Chicago* (Westport CT: Praeger).

Hackworth, J. (2005) Progressive activism in a neoliberal context: The case of efforts to retain public housing the United States, *Studies in Political Economy*, 75(Spring), pp. 29–50.

Katz, B. (2009) The origins of HOPE VI, in: H. Cisneros & L. Engdahl (Eds) *From Despair to Hope: Hope VI and the New Promise of Public Housing in American's Cities*, pp. 15–30 (Washington DC: Brookings Institution Press).

Kerlin, J. (2006) Social enterprise in the United States and Europe: Understanding and learning from the differences, *Voluntas*, 17, pp. 247–263.

Kerlin, J. (2009) *Social Enterprise: A Global Comparison* (Lebanon NH: Tufts University Press).

Kerlin, J. (2010) A comparative analysis of the global emergence of social enterprise, *Voluntas*, 21, pp. 162–179.

Lubell, J. & Barron, J. (2007) *The Importance of Integrating Rigorous Research Objectives into Any Reauthorization of the 'Moving to Work' Demonstration* (Washington DC: Center for Housing Policy).

Mullins, D. (2011) Community investment and community empowerment: The role of social housing providers in the context of 'Localism' and the 'Big Society', Third Sector Research Center Consultation Draft for HACT/TSRC Conference on Housing and Community Empowerment, June 9, 2011. Available at http://www.tsrc.ac.uk/

Mullins, D. & Pawson, H. (2010) Housing associations: Agents of policy or profits in disguise? in: D. Billis (Ed.) *Hybrid Organizations and the Third Sector*, pp. 197–218 (London: Palgrave MacMillan).

Office of the Inspector General (2004) *Audit Report: HUD's Oversight of the Philadelphia Housing Authority's Moving to Work Program* (Philadelphia, Pennsylvania, Washington DC: Department of Housing and Urban Development).

Pawson, H. & Mullins, D. (2010) *After Council Housing: Britain's New Social Landlords* (London: Palgrave MacMillan).

Smith, S. R. (2003) Governments and nonprofits in the modern age, *Society*, 40(May/June), pp. 36–45.

Smith, S. R. (2008) The challenge of strengthening nonprofits and civil society, *Public Administration Review*, 68(Special Issue), pp. 132–145.

Smith, S. R. (2010) Hyrbridization and nonprofit organizations: The governance challenge, *Policy and Science*, 29, pp. 219–229.

Schwartz, A. (2010) *Housing Policy in the United States* (New York: Routledge).

Teasdale, S. (2010) What's in a name? The construction of social enterprise, Third Sector Research Center Working Paper 46. Available at http://www.tsrc.ac.uk/

Urban Institute and Metropolitan Studies and Extended Academic Programs (2010) *A Comprehensive Affordable Housing Market Study for Mecklenburg County* (Charlotte, North Carolina: University of North Carolina Charlotte).

U.S. Department of Housing and Urban Development (2010) The Philadelphia PA, Housing Authority did not Comply with Several Significant HUD Requirement and Failed to Support Payments for Outside Legal Services, Audit Report 2011-PH-1007, March 10, 2011.

U.S. Department of Housing and Urban Development (2011a) The 1937 Housing Act. Available at http://www.hud.org

U.S. Department of Housing and Urban Development (2011b) HUD's Public Housing Program. Available at http://www.hud.org

U.S. Department of Housing and Urban Development (2011c) Amendment to the 1937 Housing Act in the 1998 Quality Housing and Work Responsibility Act. Available at http://www.hud.org

U.S Department of Housing and Urban Development (2011d) Choice Neighborhoods: History and Hope, Evidence Matters, Winter.

Utt, R. (2009) The conservative critique of HOPE VI, in: H. Cisneros & L. Engdahl (Eds) *From Despair to Hope: Hope VI and the New Promise of Public Housing in American's Cities*, pp. 249–262 (Washington DC: Brookings Institution Press).

Let a Hundred Flowers Bloom: Innovation and Diversity in Australian Not-for-Profit Housing Organisations

TONY GILMOUR* & VIVIENNE MILLIGAN**

*The Swinburne Institute, Swinburne Institute of Technology, Melbourne, Victoria, Australia, **City Futures Research Centre, University of New South Wales, Kensington, New South Wales, Australia

ABSTRACT *Australian social housing policy continues to move away from a traditional hierarchical public housing model. The small but fast growing not-for-profit sector has expanded through the introduction of private finance, a tax credit scheme, stock transfers, planning incentives and an economic stimulus package. This article examines the diverse ways in which the leading not-for-profit providers in Australia have responded to these opportunities, using the concept of organisational hybridity. Coverage of hybridity includes both established housing providers and emergent third sector organisations including finance consolidators, development consortia and cross-subsidisation vehicles. Using information from interviews, organisational case studies and documentation, this paper assesses the drivers for the growth of hybridity in Australia. The policy implications for governments steering a diverse housing sector through promoting hybrid organisations are discussed, and reflections are provided on the opportunities and limitations of using hybridity analytical frameworks. An issue to emerge from the analysis is the diversity of organisational forms, financing models and strategic orientation of hybrid organisations promoted through the same policy settings.*

Introduction

Australian governments are increasingly encouraging not-for-profit housing organisations to develop, finance and manage social housing (Milligan *et al.*, 2009). In this paper, we argue that many of the largest and fastest growing of these organisations can be conceptualised as hybrid organisations, operating between 'the respective components of the state, markets and the social capital of civil society' (Evers & Laville, 2004, p. 245). In supporting the expansion of these hybrid organisations, Australian governments are following a similar path to a number of comparable developed countries. However, while

heterogeneity among hybrid organisations is not uncommon, in Australia the diversity, innovation and pace of change is such that the outcome can be characterised as a hundred flowers blooming.

To make sense of this housing delivery transformation, this paper first reviews the policy drivers promoting hybridity. The new organisational models are then categorised using an institutional typology and illustrated with an example case study organisation for each category. Reference to international literature on hybridity provides a lens through which to assess changes in Australia. In turn, the evolving types of organisations in Australia allow reflection on the nuanced ways in which hybridity works in practice for housing providers.

Background

Australian Housing Policy Developments

Despite domestic conditions that included rapid population growth, sustained economic prosperity and an unprecedented house price boom, government investment in the net growth of social housing (across the public and not-for-profit sectors) ceased for over a decade until 2007. This set of conditions helped deepen housing affordability problems and housing stress in Australia (Yates & Milligan, 2007), triggering a newly elected social democratic government in 2007 to announce a suite of policies to support larger scale growth in the supply of housing available at below market rents (Milligan & Pinnegar, 2010). We use the term 'social housing' to encompass both deeply subsidised rental housing for those on very low income (traditionally operated as 'public' or not-for-profit 'community' housing), and 'affordable housing' for low-to-moderate income renters requiring less public subsidy. In practice, these terms are used inconsistently in Australia.

The first supply-side initiative was the 2008 National Rental Affordability Scheme (NRAS), which offers a competitively tendered funding incentive for 10 years to investors in new rental housing for low-to-moderate income households. Under NRAS, over A$4 billion of tax incentives for private investors or equivalent cash grants for not-for-profit organisations is available to part-finance a target 50 000 new rental homes (Australian Government, 2008). NRAS was followed in 2009 by an A$5.2 billion capital fund to build 19 300 new social housing dwellings intended for higher needs, lower income households (COAG [Council of Australian Governments], 2009). This 'Nation Building' project was designed to support residential construction jobs to mitigate an anticipated impact of the Global Financial Crisis.

Importantly to the focus of this article, both these supply-side initiatives have provided a major impetus to investment and development activity by not-for-profit housing organisations. This sector comprised 959 organisations till June 2010, managing just under 46 000 tenancies or around one eighth of Australian social housing, an increase from around 33 500 tenancies in 2007 (AIHW [Australian Institute of Health and Welfare], 2007, 2011). However, most not-for-profit providers remain very small, typically managing under 50 rental properties. Such small organisations do not own assets, few could raise bank borrowing and none has property development capacity. Thus their main role is to mediate between the state (as grant funder) and civil society (the communities in which they operate).

Growth and diversification in the sector is concentrated in a small number of larger organisations that have grown rapidly in scale and capacity. In 2010, 45 organisations managed 63 per cent of tenancies (Australian Government, 2010). It is these larger organisations that have benefited most from post-2008 supply-side initiatives, allowing them to grow their balance sheets and rental housing portfolios. Of the 39 292 NRAS funding incentives offered by October 2011, 55 per cent were allocated to not-for-profit providers, involving 87 recipient organisations (Australian Government, 2011). Additionally, at least 75 per cent of Nation Building social housing is being directed to larger, well-performed providers in the not-for-profit sector (Australian Government, 2010).

Other important public policy foundations for the expansion of the Australian not-for-profit housing sector include tax policy settings and tenant rental subsidy arrangements that favour not-for-profit developers over state housing authorities. Complementing these policies, several states and city councils have planning requirements or incentives for developers to provide a component of affordable housing in major new developments. Furthermore, new forms of sector regulation have recently been introduced in most states— with a promise of national regulation by 2012 (Milligan *et al.*, 2009; Travers *et al.*, 2010).

As a result of supply-side initiatives and broader policy developments, since 2008 the not-for-profit sector (rather than the public sector) has become primarily responsible for increasing the supply of social housing in Australia. Mixed financial approaches are being encouraged, with a blending of government grants, land donations, NRAS funding incentives, own resources, development profits and bank loans. Before 2008, most new development schemes were grant-funded and only a small number of not-for-profit providers used private finance or partnered with the private sector. The policy shift has led to a rapid growth in the commercialisation of the larger providers, an influx of senior managers from the private sector, the professionalisation of boards and a re-balancing between social and economic objectives.

In a relatively short period of time in Australia there has been both the emergence of new hybrid organisational forms, and the transformation of several traditional not-for-profits. It is the only country where these changes have been prompted by a move to both European-style private funding and US-style tax credits (Blessing & Gilmour, 2011). Therefore, despite challenges in making definitive assessments while in the middle of rapid change, Australia seems to be an appropriate case study country to review the drivers and organisational outcomes of moving to hybrid forms of housing delivery.

Conceptualising Hybridity

Until the 1970s, many social democratic governments including Australia delivered most social services through public sector organisations. Although the timing and extent of change has differed, from the 1980s there has been a widespread trend to more mixed forms of delivery using techniques often referred to as 'New Public Management' (Pollitt & Bouckaert, 2004). This involves a business-orientated approach by state agencies, greater partnership working with the private sector and increased delegation of service delivery to more commercialised not-for-profit organisations (Evers, 2005; Pestoff & Brandsen, 2008). Brandsen *et al.* (2005, p. 758) view 'hybridity as an inevitable and permanent characteristic' of these latter types of organisation.

Social housing is an area where new public management has had a major impact, most particularly through the increasing shift in delivery and management away from the public sector. From the mid 2000s a number of researchers have applied more general hybridity thinking to housing provision. The operation of hybrid organisations between the state, market and civil society has often been conceptualised as a triangle, with either national housing systems or individual housing organisations able to be positioned within a 'space'. For example, Mullins & Pawson (2010) conceptualise Dutch housing providers operating between the market and civil society, whereas English housing associations are more in the middle of the triangle. Other writers use a matrix approach or a linear spectrum to highlight the relative emphasis on social and economic drivers (Crossan & Til, 2009; Gruis, 2008).

The recent rise of hybridity theory in social housing research builds on an earlier tradition of institutional approaches, which emphasises the importance of understanding the organisations responsible for service delivery (DiMaggio & Powell, 1983; Powell & DiMaggio, 1991). In social housing, researchers have commented on the need to define the boundaries of the field in which housing organisations operate (Mullins & Rhodes, 2007), and the various forces that will lead different types of organisations to become more similar (Gilmour, 2009). This convergence process—also known as isomorphing—can be caused through coercive pressures such as regulation, mimetic pressures such as following role model organisations, and/or normative pressures such as setting common professional standards (DiMaggio & Powell, 1983).

Hybridity approaches to understanding the transformation of social housing organisations remain contested. A variety of conceptual models—from triangles to Venn diagrams and spectrums—have currency, with little consensus on which works best. Case study organisations seldom fit neatly within these models (Czischke *et al.*, 2010; Gilmour, 2008). As Brandsen *et al.* (2005) speculate, there may not be a rationality to hybrid organisations in terms of what they do and how they can be conceptualised. Brandsen *et al.* recommend moving beyond the study of 'safe' organisational ideal types to 'the fuzziest cases, those that can be found on the fringes of the domain' (2005, p. 762).

Purpose, Scope and Methods

This paper has two aims. First, to fill a gap in the literature by applying hybrid analytical approaches to the Australian not-for-profit housing sector. To address this, it develops an analytical typology of organisations, and discusses policy drivers that have led to hybrid housing provider diversity. The second aim is to use Australian evidence to offer insights about organisational diversity among hybrid organisations, using institutional theory that may be applicable both in Australia and other countries.

The Australian not-for-profit housing sector has until very recently been modest in scale and traditional in approach. Limited research has tracked the sector's growth through an analysis of the more innovative housing providers (Milligan *et al.*, 2004; Milligan *et al.*, 2009). There has also been research comparing Australia with international developments (Blessing & Gilmour, 2011; Gilmour, 2009; Lawson *et al.*, 2010), and looking at comparative approaches to social entrepreneurship (Blessing, 2012). No work has yet applied hybridity theory in detail.

The selected research method is to use a typological approach, which offers useful tools for classifying related items, particularly those existing in complex and fast changing environments. Typologies can help correct misconceptions and organise knowledge (Tiryakian, 1968). The approach is popular with organisational and management theorists (for example Mintzberg, 1979; Porter, 1980), and has been used to compare international trends in the not-for-profit housing sector (Czischke, 2009; Gilmour, 2009). Single case studies have been used in this paper to provide an example of each organisational type, to both illustrate and test the applicability of a typology, following the approach of Gruis (2008), who reviewed the growth of social entrepreneurism in the Dutch housing association sector.

The typology developed in this article is based on a review of secondary literature, analysis of organisations' published accounts and websites, and selected interviews with policy makers and not-for-profit housing executives. As with all typologies, the selected categories are neither exhaustive nor necessarily mutually exclusive. However the tentative categories developed are capable of inductive theory building, to be tested further in the future using a wider sample of organisations and more structured data collection (Doty & Glick, 1994).

Hybrid Housing Provider Typology

In this section we distinguish different types of not-for-profit organisations in Australia that have responded to the changing policy environment by directly investing in social housing. Organisational types are distinguished by their genesis and entry pathway into the sector, and their activities and financial approach.

The not-for-profit organisations that have been considered for classification are those larger diversified organisations described above that engage directly in investing in housing. While such hybrids only comprise an estimated 10 per cent of the not-for profit-sector, collectively they hold most of the assets that are currently owned in that sector.

Table 1 provides a comparative summary of five types of hybrid organisations, contrasting their characteristics using five 'core elements' developed by Billis (2010, pp. 48–51). These core elements have been designed by Billis to help evaluate features of organisations in the public, private and not-for-profit sectors and act as a lens through which to interpret hybridity. Later in the paper we elaborate on and consider possible reasons for differences between the social and business orientation of the five organisational types in Table 1.

The five organisational types are described below.

Entrepreneurial Traditional Not-For-Profit Housing Providers (Type 1)

Not unexpectedly at this early stage of transition to direct housing supply by the not-for-profit sector, the largest number of hybrid organisations has emerged from traditional 'community housing' providers. These were largely founded as local housing organisations to manage long-term housing funded under various government programmes from the 1980s. Typically, they began as organisations with voluntary Directors drawn from a local community and had small staff numbers. Their core activity, housing management, was funded from rents supplemented by recurrent public grants. While independent entities, they operated under contractual relationships to government with restricted autonomy and few, if any, property assets (Bissett & Milligan, 2004).

The first foray into housing development by not-for-profits in Australia came through demonstration joint venture programmes involving Type I organisations in the 1990s. Increasingly, these entrepreneurial leaders set their own strategy, engaged in mergers, expanded geographically and diversified into new activities (Milligan *et al.*, 2004; Milligan *et al.*, 2009). Type 1 organisations have followed a gradual trajectory towards exhibiting characteristics of what Billis (2010) refers to as 'embedded hybridity'.

Table 1. Comparison of Australian hybrid housing organisation types

Core elements	Type 1 *Entrepreneurial traditional*	Type 2 *Entrepreneurial welfare*	Type 3 *State-sponsored vehicles*	Type 4 *Private-sponsored vehicles*	Type 5 *Not-for-profit-sponsored vehicles*
Ownership	Typically owned by members who may comprise community agencies, residents and tenants. Generally open membership rules.	Operate either through parent body or wholly owned subsidiary of the parent. Members are usually limited to organisations affiliated with the parent agency.	Government is a founding member or share-holder. Small ordinary membership drawn from housing community typical.	Owned and managed by founding members from private sector back-grounds. May have phi-lanthropic values or faith-based affiliations. Closed membership.	Owned by founding not-for-profit organisations. Foundation member(s) have significant powers
Governance	Voluntary Directors elected by members. Recent shift to skills-based boards, with members electing Directors in expertise categories. An increas-ing share of Directors is being recruited from the private sector.	Voluntary Directors are appointed by the parent Board. Typically a mix of those with specific skills and those with links to the parent agency.	Members appoint Direc-tors under rules covering expertise and represen-tation of government. Remuneration of Direc-tors typical.	Members act as or appoint Directors who may or may not be remunerated. Most Directors have private sector backgrounds.	Member organisations appoint Directors who are typically remunerated.
Operational priorities	Business built on tenancy management and community devel-opment principles. Housing financing and development functions added as hybridity develops. Traditional low-income tenant focus, now diversifying.	Traditional welfare agency functions with housing development activities added. Some outsource tenancy man-agement. Low income and special needs tenants.	Housing development and property manage-ment core business and, in some cases, market sales. Tenancy manage-ment may be in-house or outsourced. Low and moderate income tenants.	Development and finan-cing of social housing. Tenancy management outsourced to not-for-profit or private sectors. Key worker/moderate income tenants.	Fund raising or property development/management services to member or third party not-for-profits. Tenancy management and community development done by member organis-ations. Low and moderate income tenants.

Human resources	Currently manage 500–3000 dwellings, owning a varying but growing share of these. Paid staff proportionate to scale of operation. Established in-house development and asset management capacity in recent years.	Varying scales, with up to 1000 dwellings in ownership. Small professional staff supplemented by external consultants.	Smaller than many Type 1 organisations at this early stage of development. Mix of paid staff with project management capacities and volunteers.	Smaller than many Type 1 organisations at this early stage of development. Professional paid staff for in-house functions supplemented by external consultants or partnerships with for/not-for-profit organisations.	Smaller than many Type 1 organisations at this early stage of development. Paid staff with finance, development and project management capacities.
Other resources	Historically heavily government funding reliant, today have growing revenue and asset base.	Government capital or loans on foundation.	Corporate services provided centrally by parent, which may also provide equity and share philanthropic donations. No operational government funding for housing activities.	Unpaid management time may underpin setting up. No operational funding from government.	Member contributions of start-up funding. Fee for service business model. No operational funding from government.

Source: Authors. Additional data from secondary publications, websites and annual reports till June 2010. 'Core elements' are those used by Billis (2010, p. 49).

Typically, they have strengthened their governance over the past decade by switching their Boards from comprising community representatives to being expertise-based, with a balance of social and commercial skills (Milligan *et al.*, 2009).

Most larger Type 1 not-for-profits have moved beyond their traditional mission to house low-income households paying low rents. By housing a greater mix of low-to-middle income households, organisations can increase their income stream to leverage private finance (Milligan & Pawson, 2010; Milligan & Phibbs, 2010; Victorian Auditor-General, 2010). This leveraging policy has been explicitly encouraged by governments (Australian Government, 2010).

The advent of NRAS and the Nation Building stimulus has further driven commercialisation and a move to hybrid forms by Type 1 organisations. In particular these schemes have prompted them to invest in greater in-house capacity, raise large bank loans and form new private sector partnerships. Although all five organisational types have been recipients of NRAS funding incentives, Type 1 have been most impacted. Of the NRAS funding incentives awarded to not-for-profit housing providers in the period up to October 2011, we estimate from available data that the largest share—34 per cent—has gone to Type 1 organisations, with 26 individual agencies benefiting (Australian Government, 2011). The data used here and below cover all individual not-for-profit entities with charitable status that have received an offer of NRAS funding incentives with the exception of universities that are building student housing.

Entrepreneurial Traditional Welfare Agencies (Type 2)

The second group of organisations is deeply embedded in not-for-profit traditions. However, this group has only recently diversified from aged care, disability services and homelessness service activities into providing social housing—catalysed by recent policy directions. Many in the group are large religious institutions; others are long-established diversified secular charities. Both welfare agency types have grown through philanthropy as well as government funding for service provision to be state-wide or national agencies, operating at considerable scale, in contrast to Type 1 providers, which traditionally had a more local focus. Their growth has been supported by an earlier move in Australia to outsource much social welfare provision to not-for-profit organisations, resulting in them having considerable in-house corporate capacity (Brennan & Castles, 2002).

Type 2 welfare agencies had generally already started to operate as hybrid organisations before their move into social housing provision. As with Type 1 organisations, they had adopted professional governance and diversified their revenue sources over the previous decade. Moves into the housing field by several of the well-known large welfare agencies have been stimulated by government joint venture initiatives and, more recently, NRAS funding incentives, with 25 of the Type 2 providers receiving an estimated 11 per cent of allocations across all types. As high capacity organisations with significant internal resources, this type of organisation has been in a good position to bid for and win competitive housing funds, and to recruit high-powered property senior executives.

State-Sponsored Not-For-Profit Housing Companies (Type 3)

The third type of hybrid housing organisations is a small group of state-established housing companies, similar in some aspects to the new housing associations created for the

UK's stock transfer programme. Three of these were founded in the major cities of Sydney (1994), Canberra (1998) and Brisbane (2002) as special purpose vehicles to enable government to promote arms'-length social housing development and asset management. Their business model involves using public capital or loans and public assets to leverage additional funding through cross-subsidising between commercial and social housing development. Five organisations of this type have received about seven per cent of NRAS offers in the sector (as defined above); two are among the leading national players in terms of growth and product innovation.

Unlike Type 1 organisations, these not-for-profits entities were staffed from the start by professionals with the skills and capacity to undertake commercially orientated residential developments. From their foundation they exhibited hybridity. Therefore, along with Types 4 and 5, they fall into the description of 'enacted hybrids' given by Billis (2010, p. 61). Despite using market-based development approaches, Type 3 organisations operate under a set of government mandated social aims that cannot be changed without government agreement.

Beyond the existing Type 3 cases, Australian governments have seldom pursued this model, possibly because of the subsequent successful move of traditional (Type 1 and Type 2) agencies into housing procurement (Milligan *et al.*, 2004). As the sector matures, and consistent regulation is introduced, it remains unclear whether governments will retain their strong influence over Type 3 organisations they sponsored or allow them to operate more independently. Alternatively, as foreshadowed by some (Pawson & Gilmour, 2010; Spiller & Lennon, 2009), there may be a break-up of state public housing agencies into independently governed entities similar to the Type 3 model.

Privately Sponsored Special Purpose Not-For-Profit Vehicles (Type 4)

Type 4 organisations include a small number of newly formed special purpose not-for-profit vehicles set up on commercial principles by private sector interests to finance procurement and hold housing assets. A couple of agencies of this type have also been established and skilled-up by private development firms to enable them to diversify into the urban renewal of former public housing estates. This looks set to become an expanding activity in Australia as a number of states have a policy objective of restructuring and de-concentrating public housing.

The appearance of this new organisational type is mainly linked to the introduction of NRAS, and the use of public private partnerships for housing renewal. Private sector interest in capturing financial benefits under these schemes motivates adoption of a not-for-profit entity with tax privileges. In return, the national government has been strongly supportive of NRAS funding applications from the new hybrids, with about 24 per cent of all incentives awarded to the sector up to October 2011 going to just four Type 4 organisations.

While legally Type 4 organisations are not-for-profit entities, they appear to have shallow roots in not-for-profit traditions and principles, instead having close commercial connections. Government policy allows not-for-profit housing providers to set rents at up to 75 per cent of market rates, and anecdotal evidence points to Types 4 and 5 entrants aiming towards the top of this range to achieve the required investment yields. Consequently, their tenant profiles tend to be higher rent paying 'key workers' and student groups, rather than low income and social benefit recipients.

Determining the business relationships and values of Type 4 organisations is more challenging than with the earlier provider types as they are new businesses with little track record and limited disclosure of information in the public domain. However, their continued growth appears to be tied strongly to the continuation of NRAS, which is uncertain beyond 2015.

Not-For-Profit Sector-Sponsored Special Purpose Vehicles (Type 5)

Emerging in the same context as Type 4 is another small group of special purpose organisations with similar features but sponsored by existing not-for-profits. These new organisations are either independently governed, controlled by their member organisations or have Directors in common with their parent not-for-profit. While currently having a similar legal status and business function to Type 4 organisations, motivations for their establishment seem to differ and reflect a more explicit balance of social goals and commercial imperatives.

Typically Type 5 organisations will be involved with creating critical mass by raising funds for social housing through consolidating NRAS funding incentives, or raising finance and undertaking housing development in their own right on behalf of more traditional not-for-profit housing providers. Similarly for Type 4, just three Type 5 organisations have received nearly one quarter of all NRAS offers in the sector so far. Type 5 organisations undertake relatively specialist and higher risk tasks, requiring professional staff knowledgeable in finance and property development. Many senior staff members have a private sector background, possibly being drawn to the not-for-profit sector by the motivation to achieve worthwhile social goals (personal interviews with staff).

Although Type 5 organisations are hybrid, their associated not-for-profit partners may not be. Hence establishing a Type 5 organisation can allow the more commercial activities of not-for-profit housing providers to be 'outsourced', thereby quarantining risk from the core organisation.

Analysis

In this section, common themes based on the five 'core elements' described by Billis (2010) and shown in Table 1 are identified across the organisational types, and the drivers of the growth of hybrid housing providers are assessed. Institutional theory is then used to question whether the current level of organisational diversity will continue.

Five organisations, described in Table 2, have been selected by the authors as examples that resemble the descriptions for each typological category. While examples will never be entirely typical of their type, they provide another layer of description and lend support to the analysis presented.

Comparisons Across Organisational Types

Current ownership and governance structures across the five organisational types are strongly related to the genesis and sponsorship of the agency. A few Type 1 housing providers have maintained a sizeable membership (including tenant members). This is taken by Billis (2010) to be a distinctive characteristic of third sector organisations. However, many Types 1 and 2 providers, including case study Community Housing Limited, have

Table 2. Examples of organisational types

	Type 1 *Entrepreneurial traditional*	Type 2 *Entrepreneurial welfare*	Type 3 *State-sponsored vehicles*	Type 4 *Private-sponsored vehicles*	Type 5 *Not-for-profit-sponsored vehicles*
Organisation	Community Housing Limited (CHL)	MA Housing	CHC Affordable Housing (CHC)	Ethan Affordable Housing (Ethan)	BlueCHP
Background and business model	Founded in 1994 as homelessness services provider under the leadership of the present Managing Director. CHL now largely design, develop, construct and manage social housing using grants, fees for service, NRAS, bank loans, stock transfers and retained equity.	Mission Australia (MA) is a traditional welfare agency. Founded in 1862, MA had a strong religious mission, though now largely secular. MA set up MA Housing in 2008 to provide long-term housing using NRAS, bank loans, grants, parent company equity and stock transfers.	Established by Australian Capital Territory (ACT) Government in 1998 to manage property for other housing groups, CHC moved to develop homes for sale and rent in 2000. In 2007 provided with A$50m government loan (now A$70m), used with NRAS, stock transfers and cross-subsidy from market rate developments.	Established in 2006 to support youth services. Moved in 2008 into developing social housing. With 2500 NRAS incentives, Ethan is one of the largest recipients. NRAS and investor funding used to develop new rental housing—management outsourced to private sector real estate agents.	Formed by five New South Wales traditional not-for-profits in 2008, BlueCHP develops social housing using NRAS, bank loans, founders' equity and stock transfers. BlueCHP provide procurement and property management, with tenancy management done by the five founders.
Scope	CHL operates in five Australian states and three developing countries. They manage 2000 dwellings and employ 165 staff.	MA Housing is registered in four states, employing 33 staff, and managing 1480 dwellings.	CHC operate only in the ACT, with eight staff and 229 retained rental dwellings.	Operations focussed on two states, with several development schemes underway but few properties managed as yet.	BlueCHP operates in two states, employing around 12 staff and managing 500 homes.

(Continued)

Table 2. *Continued*

	Type 1 Entrepreneurial traditional	Type 2 Entrepreneurial welfare	Type 3 State-sponsored vehicles	Type 4 Private-sponsored vehicles	Type 5 Not-for-profit-sponsored vehicles
Ownership and governance	Owned by its members, which comprise its Board of eight Directors, including the MD.	Operate through wholly owned subsidiaries of MA. Directors are senior MA executives.	Owned by members, including ACT Government. Three Directors including Chair and Deputy Chair appointed by ACT Government—the rest are skills-based, chosen by members.	Established as a not-for-profit company, thought to be owned by the three founders, who are also the sole three Directors.	A not-for-profit company owned by the five founding organisations—all not-for-profit membership organisations. Member organisations each appoint one Director.
Vision and operational priorities	CHL's vision is 'provision of affordable and sustainable housing for everyone'. They recently started employment training for disadvantaged and low-income people. CHL has partnerships across all levels of government, not-for-profit social service organisations and private firms. Strong focus on community development.	MA Housing aims to be a large-scale high-quality housing provider, facilitate access to support services, lead community development and give access to employment through social enterprises. It works with other not-for-profits to obtain support services, and partner with a for-profit NRAS developer to produce rent-to-buy housing.	CHC's mission is 'to provide affordable housing for people on low to moderate incomes' under an ACT Government stipulated 10-year growth target. Rentals are targeted to working people. Community development is not a core activity although some tenant support activities have been initiated. CHC partners mainly with private developers.	Ethan's 'vision is to see all members of the community have access to safe and affordable housing'. Most of their housing is intended for moderate income key workers (including a special scheme for police workers) and students paying 80 per cent of market rental prices. Ethan do not undertake community development work.	BlueCHP's vision is 'to maximise the availability of affordable housing in Australia through a culture of social entrepreneurship'. They provide property development and management services to third party not-for-profits on a fee basis. Community development activities are the responsibility of member organisations.

Source: Authors. Data from interviews, organisational websites and annual reports till June 2010 (quotations and metrics are from these annual reports).

moved to a position where the Board Directors are the de facto 'owners' in the absence of a genuinely participative membership structure. This also tends to be the case for Type 4 organisations on formation. However, despite the apparent concentration of power in the hands of Directors, the Boards of hybrid housing organisations have to work within a prescribed social and economic mission. This is 'policed' by a variety of stakeholders: regulators, state funding agencies and to a lesser extent staff members, the media and tenants.

Organisations across the five types are similar in that their Directors tend to come from a mix of private company, community and, to a lesser extent, public sector backgrounds. However, it is hard to precisely connect Board members' background and their outlook in their governance role. For example, Ethan has a small Board with professional skills strongly represented, yet the Directors are united by shared Christian values. Increasingly Directors, as well as the senior management team, are expected to be fully briefed across the three dimensions of hybridity: the state (public policy, grants), markets (commercial risk, private finance) and civil society (tenant needs, community integration).

The organisational missions of traditional not-for-profits (Types 1 and 2), as demonstrated by their publicly declared mission and values statements, are strongly articulated around social values, community development and social inclusion goals. Those of the newer hybrids in Table 2 are oriented more functionally towards the production of social housing and quality of service. Thus while attracted to growth and entrepreneurship, many Types 1 and 2 providers appear to be taking a more cautious approach to expanding their remit in terms of market orientation and client and product diversification. Although it can be problematic to determine an organisation's value-set from a reading of their mission statements, the types of tenants assisted reinforce the general pattern. For example, Community Housing Limited (CHL) and MA Housing have a broad mix of lower income tenants, including those with support needs. CHC Affordable Housing (CHC) and particularly Ethan are more geared towards providing rental properties for working households on low-to-moderate incomes.

In terms of operational priorities, the traditional not-for-profits (Types 1 and 2) are more vertically integrated through combining property development and maintenance, and tenancy management, than the specialist 'enacted hybrids' (Types 3, 4 and 5) set up to channel private investment. These latter organisations are more prepared to outsource tenancy management to not-for-profit or private sector companies, mirroring practice in the US, which has a tax credit scheme with some similarities to NRAS (Blessing & Gilmour, 2011; Gilmour & Milligan, 2009). Recent Australian government policy has encouraged cross-sector partnerships to finance and deliver social housing, and this has been one factor leading to a relatively low level of vertical integration of more recent hybrid housing organisations.

The majority of Australian hybrid not-for-profit housing providers carry out their activities using a mix of professional staff and outsourcing to specialist consultants. As evidenced from a selective search through housing provider annual reports, significant contributions from volunteers appear to be rare. Furthermore, while most smaller housing providers still benefit from having unpaid Board members, many larger hybrid not-for-profits have started remunerating Directors. Interestingly, some of the chief executives of the case study organisations are from private sector backgrounds (CHC), have helped manage hybrid housing organisations overseas (MA Housing) or are scholars of social enterprise (Ethan).

Billis' final category used in Table 1 is 'other resources' which in not-for-profit organisations is taken to be 'dues, donations and legacies' in contrast to taxes in the public sector and sales and fees for private organisations. This provides a good reflection of just how hybrid many Australian housing providers have become. As noted earlier, although most Type 1 (and Type 2) housing providers have always exhibited 'shallow hybridity', the clearest switch to 'entrenched hybrid' by traditional organisations has come through them seeking private finance.

Drivers of Organisational Hybridity

In part the move towards hybridity is a result of social housing funding, stock transfers and public housing regeneration projects being increasingly allocated on the basis of competitive bids. For NRAS funding incentives and stock transfers, bidders are expected to demonstrate commercial acumen and capacity in addition to housing management skills. Competitive bidding is reinforcing or developing an entrepreneurial, competitive culture in the Australian not-for-profit housing sector. This is in direct contrast to the small-scale, low-risk traditional business model anchored in community development traditions and values. Australian public policy, particularly that of the national government from 2007, has been key in this transformation. Hybrid organisations and cross-sectoral partnerships are seen as the best way to leverage scarce public capital to build more social housing, although political expectations of this strategy are exaggerated (Milligan & Pawson, 2010). This shift is occurring in response to widespread acceptance that the public sector has been unable (or unwilling) to provide sufficient social housing (Blessing, 2012).

Government policy, therefore, is key to understanding the decisive shift to the use of hybrid organisations for delivering growth in Australian social housing. NRAS and Nation Building together provided the highest level of direct public investment in social housing since the 1980s, enabling hybrid organisations to grow quickly. However, the seeds of hybridity were sown by precursors overseas, especially the successful development of a variety of forms of housing associations in Britain and the Netherlands from the 1970s. The working of these overseas examples was in part demonstrated by the handful of Australian entrepreneurial Types 1 and 3 organisations, which adopted new entrepreneurial directions from the early 1990s (Bissett & Milligan, 2004; Milligan *et al.*, 2004). It was the success—albeit on a small scale—of these organisations and projects that facilitated greater local political support for such approaches from 2007.

Although over the past three decades it has largely been governments that have prompted the much greater use of hybrid approaches to housing delivery, through moves to introduce external funding, the organisational outcomes will depend on the choice of financial levers used. Where bank lending is the main lever, existing housing providers will need to improve their professionalisation through disciplines imposed through the banker–customer relationship. By contrast, where tax credit-supported equity is the main lever selected by government to bring market forces to social housing delivery, more special purpose hybrid vehicles and intermediate institutional forms, such as tax credit consolidators will be formed (Gilmour, 2009).

One of the main reasons for the innovation and diversity seen among contemporary Australian not-for-profit housing providers is that government has attempted to stimulate the use of both private debt *and* equity financial levers. The organisational outcomes, therefore, share some similarities with both Britain and the US. NRAS funding incentives, like US housing tax credits, are contested between for-profit and not-for-profit bidders.

Income can be earned by organisations packaging tax credits, leading to a growth of intermediation vehicles such as some Types 4 and 5 organisations.

Institutional Theory and the Future of Organisational Diversity

The tendency of organisations operating in the same sector (organisational field) to become more similar through processes of coercive, mimetic and normative isomorphism was discussed earlier in this paper. Coercive isomorphism in the Australian sector through regulation has to date been limited, with one of the few impacts being the general move of Type 1 hybrid housing providers from association or co-operative status to register under Corporations Act and the formation of new hybrids similarly. The broader impact of sector regulation on conforming organisational approaches to governance and operational priorities has been modest in Australia as specialised regulation has only been recently introduced, is restricted to certain states, differs between jurisdictions and there is as yet no national approach (Travers *et al.*, 2010, 2011).

There is little evidence of mimetic isomorphism occurring *across* the organisational types, where particular housing providers are held up as role models for others to follow (although there is some evidence of this occurring *within* the Type 1 category where there are acknowledged 'sector leaders'). Furthermore, normative isomorphism is limited as the social housing professional staff body has a limited impact, and the trade associations focus on provider Types 1, 2 and 3 (Gilmour, 2009).

Although the current policy and organisational background in Australian has encouraged a diversity of hybrid types, this may not continue. Forces promoting coercive isomorphism look set to strengthen, with a promised move to consistent national regulation of most not-for-profit housing providers by June 2012 (Arbib, 2011). National regulation will probably encompass all organisational types except perhaps Type 4, though case study Ethan Affordable Housing has indicated that they are considering registration in any event (personal communication). Furthermore, normative isomorphism may become more of a force with common reporting required for NRAS, similar control procedures for bank borrowing and a continuing influx of senior staff from the private sector.

Conclusions

Despite a common move in several countries towards hybrid not-for-profit models of housing production, local organisational outcomes will depend on a range of factors, particularly the finance levers selected for introducing external finance and the policy, institutional and regulatory settings.

Using recent developments in Australia as the test bed, this paper has offered a typology that can form the basis of future research both in Australia and elsewhere. Five distinctive types of hybrid organisations operating within the not-for-profit sector have been identified: traditional housing and welfare organisations that have adopted an entrepreneurial ethos, as well as an array of newer state, private or not-for-profit-sponsored special purpose vehicles. The hybrid diversity found in Australia can be explained in part through different policy contexts that operated at the time individual organisations were established and by the influence of varying policies at state government level. The most recent influential policy shift has been the move to private finance in the form of both debt finance and tax-credit incentives, with the latter creating a need for (and lucrative source of fee income for) professional intermediaries. In addition, the Australian Government has set up an explicit policy objective of professionalising the existing not-for-profit housing sector through commercial disciplines and private sector partnerships.

In building the typology, the 'core elements' approach by Billis (2010) to understanding cross-sector organisational approaches, and the concept of 'enacted hybrids', have been valuable. Given the very recent move to 'entrenched hybridity' in Australia, later studies will be able to comment more authoritatively on whether the identified trends continue. In particular there will be greater clarity on the rate of growth, operational priorities and governance of the 'enacted hybrids' (Types 3, 4 and 5). The typology has also demonstrated the distinctly different possible forms of 'enacted hybrids', and examples from Australia (especially of Type 3 organisations) show how they can develop strategies and funding approaches that were probably unintended by their founders. Furthermore, with the move away from membership style structures, several case study hybrid organisations have governance structures that blur separation between managers, directors and owners and may result in weaker external accountability. There are also public policy issues with special purpose vehicles (Types 4 and 5) apparently established to mitigate tax payable on commercialised activities. This raises questions on whether charity laws are being used appropriately (Blessing, 2010).

Although this research is focused on one country, there are a number of issues emerging that should resonate with other jurisdictions using hybrid organisations to deliver social housing policy. The hybrid housing organisations examined in this article face the dilemmas implicit in balancing social and economic missions, in particular there is a risk of them 'becoming too attached to commercial motivations and losing touch with their social objectives' (Gruis, 2008, p. 1091). Some of the more recently formed special purpose vehicles in Australia (Types 3 to 5) have relatively shallow community anchorage, an issue highlighted by Mullins (2006) in his research on the English move to hybrid-style housing organisations. These issues are also true for several of the traditional Australian housing providers (Types 1 and 2) that have moved to a more business-like model. Interestingly in this context, it is a handful of Type 1 providers that has been at the forefront in Australia of moving from a regional to a national housing delivery model.

Blessing (2012) positions the Australian government as promoting not-for-profit housing organisations as a 'magical solution' to entrenched housing policy problems through harnessing their 'hybrid vigour'. By contrast in the Netherlands the hybrid housing associations are now seen as 'monstrous'—criticised for high executive salaries, reckless risk-taking and lacking community focus. The typological approach developed in this paper draws attention to differences *between* hybrid housing provider types. For example, an apparent split emerges between the more traditional organisations (Types 1 and 2), which retain a balanced social and economic mission, and some of the Types 3, 4 and 5 housing providers with less grounded social missions and riskier business models.

Widespread diversity among Australian not-for-profit housing organisations may not necessarily have been the outcome desired by government, and may moderate the speed at which the sector can grow. For example, organisational heterogeneity limits the ease of staff and knowledge transfers, and minimises economies of scale for professional service providers, lenders and tax-credit investors. Using the lens of institutional theory, the Australian not-for-profit sector currently has a low level of institutionalisation, with limited forces promoting coercive isomorphism.

The plurality encouraged by Chairman Mao Zedong's promise to let a hundred flowers bloom in early 1957 did not last long. Soon the Chinese state re-asserted control, enforcing conformity. Similarly in contemporary Australia, the extraordinary diversity of, and innovation by, housing providers of the last four years may be short-lived. Institutional factors such as the proposed introduction of national regulation will most likely encourage greater convergence in the *modus operandi* of different housing provider types. Furthermore, unless more sustainable public policy settings and funding strategies emerge, the shelf life of high-risk not-for-profit business models may be short-lived. Perhaps not all the new not-for-profit organisational types recently emerging in Australia are set to become hardy perennials.

References

AIHW (Australian Institute of Health and Welfare) (2007) *Community Housing 2006–07. Commonwealth State Housing Agreement National Data Report* (Canberra: Australian Institute of Health and Welfare).

AIHW (2011) *Community Housing 2009–10* (Canberra: AIHW).

Arbib, M. (2011) Senator Arbib's Address to the Power Housing Conference, Melbourne, March 16.

Australian Government (2008) *National Rental Affordability Scheme Regulations 2008: Select Legislative Instrument 2008 no. 232* (Canberra: Australian Government).

Australian Government (2010) *Regulation and Growth of the Not-For-Profit Housing Sector: Discussion Paper* (Canberra: Australian Government).

Australian Government (2011) *National Rental Affordability Scheme Monthly Performance Report 31 October 2011* (Canberra: Australian Government).

Billis, D. (Ed.) (2010) *Hybrid Organisations in the Third Sector: Challenges of Practice, Policy and Theory* (Basingstoke: Palgrave).

Bissett, H. & Milligan, V. (2004) *) Risk Management in Community Housing. Managing the Challenges Posed by Growth in the Provision of Affordable Housing* (Sydney: National Community Housing Forum).

Blessing, A. (2010) Public, Private or In Between? Defining the Legal Status of Non-Profit Housing Providers. Paper presented at the European Network for Housing Reasearchers' Conference, Istanbul, July 2010.

Blessing, A. (2012) Magical or monstrous? Hybridity in social housing governance, *Housing Studies*, 27(2), pp. 189–207.

Blessing, A. & Gilmour, T. (2011) The invisible hand? Using tax incentives to encourage institutional investment in social housing, *International Journal of Housing Policy*, 11(4), pp. 453–468.

Brandsen, T., van de Donk, W. & Putters, K. (2005) Griffins or chameleons? Hybridity as a permanent and inevitable characteristic of the third sector, *International Journal of Public Administration*, 28, pp. 749–765.

Brennan, G. & Castles, F. G. (Eds) (2002) *Australia Reshaped: 200 Years of Institutional Transformation* (Cambridge/Port Melbourne: Cambridge University Press).

COAG (Council of Australian Governments) (2009) *National Partnership Agreement on the Nation Building and Jobs Plan: Building Prosperity for the Future and Supporting Jobs Now*. Council of Australian Governments (COAG). Available at www.coag.gov.au (accessed 4 May 2011).

Crossan, D. & Til, J. (2009) Towards a Classification Framework for Not-For-Profit Organisations: The Importance of Measurement Indicators. EMES Conferences Selected Papers. Available at http://www.emes.net/fileadmin/emes/PDF_files/Selected_Papers/Serie_1_Theme_1/ECSP-B08-01__Crossan-Van_Til_.pdf (accessed April 6 2012).

Czischke, D. (2009) Social Entrepreneurship in Housing: Exploring Missions, Values and Activities through Case Studies in Three European Countries. Paper presented at the ENHR Conference, Prague, 28th June–1st July.

Czischke, D., Gruis, V. & Mullins, D. (2010) Conceptualizing Social Enterprise in Housing Organisations. Paper presented at the European Netwok for Housing Research Conference, Istanbul, July.

DiMaggio, P. J. & Powell, W. W. (1983) The iron cage revisited: Institutional isomorphism and collective rationality in organisational fields, *American Sociological Review*, 48(2), pp. 147–160.

Doty, D. H. & Glick, W. H. (1994) Typologies as a unique form of theory building: Toward improved understanding and modelling, *Academy of Management Review*, 19(2), pp. 230–251.

Evers, A. (2005) Mixed welfare systems and hybrid organizations: Changes in the governance and provision of social services, *International Journal of Public Administration*, 28(9), pp. 737–748.

Evers, A. & Laville, J.-L. (2004) Social services by social enterprises: On the possible contributions of hybrid organizations and a civil society, in: A. Evers & J. -L. Laville (Eds) *The Third Sector in Europe*, pp. 237–255 (Cheltenham: Edward Elgar).

Gilmour, T. (2008) Same or different? Towards a typology of nonprofit housing organisations, in: A. Jones, T. Seelig & A. Thompson (Eds) *Reshaping Australasian Housing Research: Refereed Conference Papers and Presentations from the 2nd Australasian Housing Researchers' Conference* (Brisbane: University of Queensland).

Gilmour, T. (2009) Network power: An international study of strengthening housing association capacity. Available at www.ahuri.edu.au (accessed 4 May 2011).

Gilmour, T. & Milligan, V. (2009) Stimulating institutional investment in affordable housing in Australia. Insights from the US, in: T. Dalton, V. Colic-Peisker, E. Taylor & R. Muir (Eds) *Refereed Papers Presented at the 3rd Australasian Housing Researchers Conference, Melbourne, June 18– 20 2008* (Melbourne: RMIT University).

Gruis, V. (2008) Organisational archetypes for Dutch housing associations, *Environment and Planning C: Government and Policy*, 26(6), pp. 1077–1092.

Lawson, J., Gilmour, T. & Milligan, V. (2010) *International Measures to Channel Investment Towards Affordable Rental Housing* (Melbourne: Australian Housing and Urban Research Institute [AHURI]).

Milligan, V., Gurran, N., Lawson, J., Phibbs, P. & Phillips, R. (2009) *Innovation in Affordable Housing in Australia: Bringing Policy and Practice for Not-For-Profit Housing Organisations Together: Final Report 134* (Melbourne: AHURI).

Milligan, V. & Pawson, H. (2010) Transforming Social Housing in Australia: Challenges and Options, Paper presented at the European Network for Housing Reasearchers' Conference, Istanbul, July. Available at http://enhr2010.com/fileadmin/templates/ENHR2010_papers_web/papers_web/WS07/Ws07_208_ Milligan.pdf (accessed 11 May 2011).

Milligan, V. & Phibbs, P. (2010) Building a not-for-profit affordable housing industry in Australia, in: B. Randolph, T. Burke, K. Hulse & V. Milligan (Eds) *Refereed Papers Presented at the 4th Australasian Housing Researchers Conference, Sydney, August 2009* (Sydney: City Futures Research Centre, University of New South Wales).

Milligan, V., Phibbs, P., Fagan, K. & Gurran, N. (2004) *A Practical Framework for Expanding Affordable Housing Services in Australia: Learning from Experience: Final Report 65* (Melbourne: AHURI).

Milligan, V. & Pinnegar, S. (2010) The comeback of national housing policy: First reflections, *International Journal of Housing Policy*, 10(3), pp. 325–344.

Mintzberg, H. (1979) *The Structuring of Organizations: A Synthesis of the Research* (Englewood Cliffs, NJ: Prentice-Hall).

Mullins, D. (2006) Competing institutional logics? Local accountability and scale and efficiency in an expanding non-profit housing sector, *Public Policy and Administration*, 21(3), pp. 6–24.

Mullins, D. & Pawson, H. (2010) Hybrid organisations in social housing: Agents of policy or profits in disguise? in: D. Billis (Ed.) *Hybrid Organisations in the Third Sector: Challenges of Practice, Policy and Theory*, pp. 197–218 (Basingstoke: Palgrave).

Mullins, D. & Rhodes, M. L. (2007) Special issue on network theory and social housing, *Housing, Theory and Society*, 24(1), pp. 1–13.

Pawson, H. & Gilmour, T. (2010) Transforming Australia's social housing: Pointers from the British stock transfer experience, *Urban Policy and Research*, 28(3), pp. 241–260.

Pestoff, V. & Brandsen, T. (Eds) (2008) *Co-Production. The Third Sector and the Delivery of Public Services* (Abingdon: Routledge).

Pollitt, C. & Bouckaert, G. (2004) *Public Management Reform: A Comparative Analysis* (Oxford: Oxford University Press).

Porter, M. E. (1980) *Competitive Strategy: Techniques for Analyzing Industries and Competitors* (New York: Free Press).

Powell, W. W. & DiMaggio, P. J. (Eds) (1991) *The New Institutionalism in Organizational Analysis* (Chicago, IL: University of Chicago Press).

Spiller, M. & Lennon, M. (2009) Re-inventing social housing, a once in a generation chance, *HousingWorks*, 7(1), pp. 20–22.

Tiryakian, E. A. (1968) Typologies, in: D. L. Sills (Ed.) *International Encyclopaedia of the Social Sciences* (New York: Macmillan).

Travers, M., Gilmour, T., Jacobs, K., Milligan, V. & Phillips, R. (2011) *Stakeholder Views on the Regulation of Affordable Housing Providers in Australia: Final Report 161* (Melbourne: AHURI).

Travers, M., Phillips, R., Milligan, V. & Gilmour, T. (2010) *Regulatory Frameworks and Their Utility for the Not-For-Profit Housing Sector: Positioning Paper 127* (Melbourne: AHURI).

Victorian Auditor-General (2010) *Access to Social Housing* (Melbourne: Victorian Auditor-General).

Yates, J. & Milligan, V. (2007) *Housing Affordability: A 21st Century Problem: Final Report 109* (Melbourne: AHURI).

Expansion, Diversification, and Hybridization in Korean Public Housing

HYUNJEONG LEE* & RICHARD RONALD**, ***
*Department of Housing and Interior Design, Kyung Hee University, Seoul, South Korea,
**Department of Geography, Centre for Urban Studies, Planning and International Development Studies,
University of Amsterdam, Amsterdam, The Netherlands, ***Department of Housing and Interior Design,
Kyung Hee University, Seoul, South Korea

ABSTRACT *The emphasis in European contexts has been on the residualization and market orientation of social housing agencies. South Korea has, however, experienced an extension of public housing and greater sector diversification around different forms of provision that serve the needs of various types of household. On the one hand, a permanent public rental housing sector has been developed serving the needs of very low-income or vulnerable households. On the other hand, more diverse types of fixed-term rental have been produced for a broader range of income categories. This paper examines how and why these differentiated approaches to public housing provision and management have emerged. A particular focus is the changing demands on and roles of housing organizations, as well as the relationships between government, market, and civil sector organizations. The analysis identifies a particular hybridity in South Korean public housing in contrast to typologies developed in European contexts.*

Introduction

While in European societies the residualization of social housing sectors and the transformation of social providers into more market-orientated or socially entrepreneurial entities have been characteristic (Cowans & Maclennan, 2008; Priemus & Dieleman, 2002; Rhodes & Mullins, 2009), South Korea has followed a particular path in which public forms of social housing have been diversified and extended to serve greater numbers and types of households (Ronald & Lee, 2012). This paper considers recent institutional transformations in the Korean social housing sector. A permanent public rental housing sector has been developed since 1989 serving the needs of very low-income households (Ha, 2004; Lee & Hong, 2007; Lim, 2005; Park, 2007), although, more recently, more diverse categories of public housing have been produced for low- and middle-income households. Different sectors have demanded distinctive approaches to provision and management, and while permanent public rental housing has become more embedded in social partnerships with local government and civil sector agencies, and focused on serving residents' needs, public housing support for middle-income households has needed to reconcile the need for affordable housing with commercial practices and market contexts.

The development of these qualitatively different public housing sectors has primarily been the responsibility of the Korea Land and Housing Corporation (LH), founded in 2009 with the merging the Korea National Housing Corporation (KNHC) with the Korea Land Corporation. Public housing coordination in (South) Korea has changed significantly over the last two decades reflecting changing state approaches and shifting housing, demographic, and socioeconomic conditions. The LH has been required to both develop different versions of the social housing sector and support stability in the private housing market. These different demands have, as well as taking a considerable toll on LH resources, raised questions about the role of housing policy and relationships between private, public, and civic sectors in the provision of housing and welfare.

The development of (South) Korean housing has followed a distinct path shaped by the emergence of 'developmental' forms of government during a prolonged period of economic expansion (Park, 1998). Since the late 1990s, however, there have been marked shifts in economic growth, democratization, along with neoliberal transformations in regulation and patterns of capital accumulation, especially in real estate. This context has shaped very different institutional responses, especially in the social housing sector. This paper, through the examination of elaboration, transformation, and hybridization in Korean social housing provision, seeks to contribute to the comparative understanding of the dynamic evolution of housing institutions and policy frameworks. The Korean case illustrates considerable diversity in how, and under what conditions, constellations of social and public housing providers, managers, and users come together and interact. These challenge assumptions drawn in Western contexts concerning relationships between state, market, and private organizations.

The first part of this paper addresses prevailing conceptual frameworks regarding the emergence and transformation of social housing providers. It goes on to contrast these with East Asian cases, identifying important differences in the logic and organization of public housing. The second part considers the development of social housing in South Korea and, in particular, the extraordinary elaboration of social housing programmes in the last two decades. We consider what emerging forms of public and social housing, and examine how institutions have been transformed. The role of state agencies, like the LH, has shifted from facilitating private development to active provision of social housing services. The public housing sector has, however, become complex, requiring different institutional constellations pursuing different goals in different spheres of housing provision. The last section addresses the most recent era of policy innovation, changing demands in housing provision and the significance of the Korean case for understanding processes of hybridization in social housing.

Housing Systems and Organizations

Typologies and Transformations

Understanding trends in social housing sectors and housing organizations has been dominated by European examples reflecting their historic embeddedness in this context.

Social housing sectors across Europe are characteristically diverse in terms of sector size, legal and organizational forms of social housing providers (from public bodies to cooperatives and not-for-profit organizations), forms of social tenures (renting and affordable ownership, etc.), and the overarching housing policy frameworks under which providers operate (Czischke, 2009, p. 126). It is, nonetheless, possible to distinguish between two key types based upon Kemeny's (1995) model. *Unitary* systems stem from a particular conception of social welfare that aims to provide housing of decent quality at affordable prices for the population as a whole. Within this approach, housing is typically considered a public responsibility and delivered through municipal housing or nonprofit organizations. The other type is a *dualist* system, which assumes that housing needs will largely be met by the market. Social housing interventions are thus extended only to those on the margins of society for whom the market is unable to deliver decent or affordable housing.

Starting in the late 1970s, but becoming more intensive in the last two decades, transformations in social housing sectors have, however, been radical. In line with neoliberalization more broadly, market concepts and market forms of coordination have been in the ascendancy in housing policy and among housing organizations in both unitary and dualist contexts. Interventions have driven privatization and residualization of the sector, representing a shift away from the provision of housing as a collective good delivered by special purpose public bodies (Malpass & Victory, 2010; Rhodes & Mullins, 2009). Increasingly, regulations have also been imposed on housing associations concerning how much subsidized rental housing they should provide and the range of low-income households that should qualify (Priemus & Gruis, 2011).

Despite the trend away from government and hierarchical control toward the reinforcement of market principles and contractual relationships, transformations in provision and management of social housing in each country retain considerable path dependency, reflecting not only pre-existing scales of provision and frameworks of regulation and distribution, but also the uneven impact of market principles on institutional relationships. Following the ascendency of neoliberalized approaches to social housing, Ghekiére (2007) distinguishes two further subtypes that represent more targeted versions of provision. In one, housing is allocated in terms of those falling under an income ceiling (*generalist*), and in the other, social housing provision is for the most vulnerable (*residual*). While the former follows a more traditional conception of 'social' housing, albeit to a more restricted population, the latter caters for a more restricted category of beneficiary.

Restructuring has helped confirm a status change in social rental housing from a public good for all citizens to a contingent service for the very poor, excluded or vulnerable. It has also been accompanied by a sharp reorientation in the organizations that build and manage social housing. This has involved a substantial shift away from public ownership and control at the macro-level. At the micro-level, transformations have required internal structural reform, increasing interaction with external private sector organizations (through partnerships, procurement or competition) and the management of these relationships through competition-based regulations (Rhodes & Mullins, 2009, p. 112). Despite greater market orientation, the withdrawal of direct state welfare provision has brought increasing pressures to bear on social landlords to become more 'socially entrepreneurial', providing social services such as community development, employment generation, and youth projects (Czischke *et al.*, 2010).

Many European housing providers have effectively been transformed into *hybrid* organizations that combine features of the third sector, state, and commercial organizations with competing social and market logics. Such hybridity is characteristically triangular involving state, community, and market angles intersected by public–private, formal– informal, and profit–nonprofit axes (Brandsen *et al.*, 2006). Walker (2001) identifies changes in social housing with 'new public management' in which organizations have moved from an environment of comfort to competition by adopting reforms involving disaggregation, competition, private sector practice, and performance management. The rest of this paper addresses developments in, and considers what hybridity means for social housing providers in developed East Asia where responses to changing conditions have been remarkably converse. South Korea provides a particularly illustrative example as housing providers have also faced the combined challenges of serving community, welfare, and market roles. In this context, however, social housing provision, along with other social welfare policies, has been rolled out rather than retracted.

Developmentalism as a Context for Public Housing Provision

While East Asian economies have diverse and usually limited public programmes supporting the welfare of citizens, most countries have, since the 1960s, engaged in large-scale public housing programmes (Chiu, 2008). East Asian political relations have typically involved powerful political elites forming strong alliances with state bureaucracies and the managers of large private companies with the primary objective of coordinating economic growth. These so-called 'developmental states' have featured authoritarian governments, top-down planning, and state management with the premise of deep public interventions being the development of the private market and corporate growth (Wade, 1990). Large-scale public housing interventions have thus primarily sought to serve *developmental* goals: to clear slums, house workers, extend urban infrastructure, capture potential land values, and sustain employment and the rapid growth of the economy more generally.

In this context, very large public organizations have been established to coordinate housing production and consumption activities. Examples include the Hong Kong Housing Authority (HKHA), Singapore's Housing Development Board (HDB), Japan's Urban Renaissance Agency, and Korea's LH. The power wielded by such organizations has been considerable. Through the ownership and control of land and mass home building, such organizations have been able to transform housing sectors. Most notable are the HKHA and the HDB, as while the former provides housing for around 48 per cent of Hong Kong's 7-million-plus inhabitants, the latter has, since the late 1960s, built more than 83 per cent of the housing stock for Singapore's 4-million population.

In this context, *public housing* in East Asia can be distinguished from European *social housing* (Ronald & Lee, 2012). Firstly, public housing has not acted to de-commodify the housing market, but rather sustain its growth and development. Homes produced by state agencies are often sold to households as a form of owner-occupied housing, or, where produced as subsidized rental dwellings, have not been allocated to the neediest, but rather to productive categories of worker to support economic development objectives. For example, subsidized state flats have often been considered a form of social wages that allow companies to keep salaries low making industry more competitive (Castells *et al.*, 1990).

Secondly, through large public housing interventions governments have sought to demonstrate competency and shore up electoral success (Chua, 1997). An important contrast with European contexts is that the state and public agencies have—until the late 1990s at least—been considered successful in, if not essential to, the delivery and development of modern housing stock and real estate-driven capital accumulation among growing numbers of middle-class households. Governments have planned, coordinated, and supported large-scale housing projects, albeit on, typically, market terms (Doling, 1999). Public housing has thus not represented an obstacle to capital, nor government intervention a contradiction to a market economy. Indeed, state management has not been considered an impediment to economic advancement, as in European policy discourses, but rather a requirement.

Thirdly, public housing has shaped the development of East Asian welfare. Social policy in developmental countries has typically been subordinated to the requirements of intensive economic growth (see Kwon, 2009), meaning welfare state expansion has been either sacrificed or focused in ways that contribute to economic objectives or reinforce the commodification of labor. Public housing plays a particular role in this regard, with allocation targeting the economically active and housing often provided in commodified forms, such as owner-occupied dwellings that accrue in asset value and can enhance the capacity of family self-reliance (Ronald, 2007). Public housing thus represents a means to offset social welfare spending while focusing investment in infrastructure and supporting economic growth.

Since the late 1990s, however, there have been marked changes in socioeconomic conditions and public housing approaches. Following the 1997 Asian Financial Crisis, the region entered a new era in which high-speed economic growth and full employment no longer appears assured. The initial collapse of property values revealed that state housing interventions were misaligned with emerging demand conditions (Chiu, 2008). In the postcrisis decade, markets became characteristically volatile, forcing many developmental regimes to reassess policy strategies and interventions.

For Gill & Kharas (2009), 'growth-first' policies have been tempered by new strategies that assume that market forces must be managed, not just to generate growth, but also to avert market failure. Developmental states have become more concerned with guiding economies by providing an environment in which economies of scale can flourish while also managing the distributional outcomes and socioeconomic polarization generated by such an economy. This shift has, furthermore, unfolded along with increased political contestation. The crisis undermined the legitimacy of past policies, sustaining a trend toward greater democratization. The crisis exposed the inadequacy and social inequity of developmentalist era welfare arrangements as when the economy failed there were few social safety nets to fall back on. As countries like Korea had already achieved high GDP per capita levels, there were expectations that the state could now do more and social policy became an arena in which political parties contested power, especially as differences between rich and poor began to grow again (Peng, 2004).

Public Housing Transformations in South Korea

While many East Asian countries have subsequently rolled out social policies and reinforced social forms of public housing (Ronald & Chiu, 2010), South Korea has arguably been the most prominent. The rest of the paper examines Korean experiences beginning with early initiatives that reflected developmentalist objectives. These advanced the production of housing stock, but without developing a consistent provisionary framework for low-income households. The analysis goes on to consider the hybridized forms of public housing that emerged after 1989 and their development through the 1990s and 2000s. Attention then turns to the present institutional structure as well as how it is reacting to, and being reformed by the latest crisis. Ultimately, the meaningfulness of European concepts is considered in terms of the ongoing hybridization of public housing and conflicting demands placed on social providers.

Early Public Housing Interventions

While the KNHC was formed in 1962 as a self-financing public agency, public housing was legally established in the 1963 Public Housing Act. This Act identified the KNHC and local governments as responsible for building housing for low-income households. Nonetheless, due to funding constraints the KNHC primarily built owner-occupied units that facilitated quicker cost recovery. Local governments, meanwhile, were largely left with the task of providing housing for households unable to afford to buy a home in the private market, with urban migration placing severe pressure on housing conditions.

The Housing Funding Act of 1963 designated national funding for housing production in both public and private sectors. The Korea Housing Bank (KHB) was later established in 1967 and, after the Housing Construction Promotion Act of 1972, made loans to both public entities and private housing development companies to build homes. Despite the emphasis on public housing production, a dual system emerged in which private interests dominated. Although land, finance, and tax benefits existed, incentives and resources for the construction of housing for the poor were inadequate.

Within the national '10-year Housing Construction Plan', started in 1972, the first KNHC public rental housing scheme for low-income people was initiated. Housing was provided with tenancy limited to 1 year and the property sold thereafter to, in principle, tenants. Short-term tenancies meant that the rental housing program was as much an assisted home purchase scheme that facilitated the quick recovery of costs for the KNHC (Lee & Hong, 2007). Supply of this tenure was limited, representing only 2–3 per cent of total housing output over this period. In 1982, a form of 5-year public rental tenancy housing targeting middle-income working households was introduced and later expanded among private suppliers under the Rental Housing Construction Promotion Act of 1984.

The rising democratic and civil rights movements also pushed localism in the 1980s leading to the setting up, and later expansion of local public housing agencies across South Korea's seven metropolises, including the capital, and nine provinces. Through these 16 public housing agencies, local governments aimed to provide rental and owner-occupied housing for lower-income households. As this was primarily achieved through the implementation of urban clearance, agencies were often called Urban Development Corporations.[1]

Inevitably, the measures adopted before 1989, while establishing a legal and institutional framework for public housing provision, did relatively little to provide housing for the poor. Low-income households were typically forced to turn to the market while government subsidies flowed into supply initiatives, assuming the more housing was constructed, the quicker shortages would be resolved (see Park, 2007; Ronald & Jin, 2010). Few large developers were interested in building rental housing, especially not for low-income households (Lim, 2005). Demand-side subsidies were unexplored although government agencies began to supply *Chonsei* (or *Jeonsei*) loans for households taking on short-term leases.[2] In fact, the proportion of *Chonsei* to housing tenure expanded from 17.3 per cent of housing in 1975 to 29.7 per cent by 1995 (KNHC, 2005), as a result of the flow of urban housing for sale into the hands of investors. In context of high demand for rental housing, but little provision, pressure to provide for the poor and socially vulnerable grew.

Public Rental Housing Programmes Become More Social

In the late 1980s, in context of increasing democratization and amplified demands for more adequate and affordable housing, pressures increased on the KNHC to become more proactive in supporting low-income rental households. Meanwhile, other institutional initiatives served to promote housing provision for working households, but usually within the market framework. The National Housing Fund (NHF) was established in 1981 and increasingly supported various modes of housing provision and consumption, including urban regeneration, supply-based rental, and for-sale housing programmes as well as first-home buyer's loans. The fund was administered by the KHB and derived from sources such as National Housing Bonds, deposits from the Housing Subscription Scheme (HSS), direct subsidy from the government, lottery funds, and others. Figure 1 illustrates the diverse sources of NHF revenue. While the NHF was established to fund low-income housing, it increasingly provided supplementary funding as the government propensity became more direct intervention through the KNHC. The changing nature and funding of low-income rental housing shaped the development of three qualitatively different public rental housing sectors in the 1990s: Permanent Public Rental Housing, Long-term Public Rental Housing (LT-PRH), and National Public Rental Housing (NPRH).

The Permanent Public Rental Housing (PPRH) program is arguably the first and possibly only true social rental housing program introduced in Korea. This tenure was intended to provide 250 000 units with indefinite tenancy for households in the lowest-income decile. It initially derived 85 per cent of its funding from the government and 15 per cent from NHF finance (KMLTMA, 2008; STFSAP, 2007). The housing itself was produced and managed by the KNHC, making the role of social landlord one of its institutional responsibilities, to be managed alongside support given to the private construction of commodity housing. In the end, 190 077 PPRH units were constructed from 1989 to 1993. Production ceased

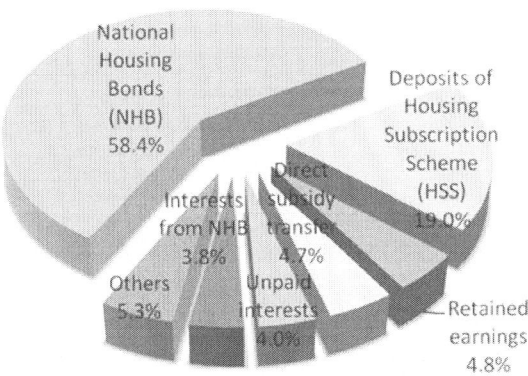

Figure 1. Distribution of revenue sources of the National Housing Fund. *Note:* Total NHF revenue in 2010 was 83.9-trillion KRW. *Source*: KMLTMA (2011b).

earlier than planned due to miscalculations in projecting the number of tenants and burdens on government spending.[3] The eligibility of households for PPRH has shifted emphasis over time, and while the initial programme targeted the lowest-income decile of households, the focus has become specific including socially vulnerable categories such as recipients of Basic Livelihood Security, single parents, North Korean refugees, poor elderly, and disabled households. Essentially, developments in this sector illustrate some shift from a generalist to a residual model of social housing provision, although such an interpretation must account for expansion in other forms of public rental provision.

Long-term, fixed period public rental housing has been constructed for lower-income households above the income threshold of PPRH. The LT-PRH replaced PPRH construction, helping reduce the financial burden of public rental housing production. The state contribution to construction costs was 50 per cent, with 20 per cent from the NHF, 10 per cent from tenants' deposits and the rest (10 per cent) derived from developers. Between 1992 and 1997, 92 730 small units sized were built. Housing was constructed by the KNHC and local housing agencies (LHA) on behalf of central and local governments. Allocation has prioritized people displaced by urban development or regeneration projects, people with 'distinguished service to the nation' and low-income HSS account holders (KMLTMA, 2008). This kind of tenure has a fixed period (50 years) during which the dwelling can be designated for social renting. LT-PRH for low-income groups contrasts to Short-Term Rented Housing (STRH) programmes provided for middle-income households, with STRH tenants expected to buy their home after a predetermined 5-year period. Different forms of STRH include Public- (from 1993), Workers- (1989–2004) and Company (from 1982) 5-year rental, built by employers and private developers as well as KNHC and LHA.

The NPRH program was introduced in the early 2000s and aimed to provide for a broader class of working households. Government subsidy accounts for up to 40 per cent of total construction costs, so much higher deposits are required. The maximum rental tenure is fixed at 30 years. Although the poorest 40 per cent of Korean households theoretically qualify for NPRH, higher user costs mean that households with moderate incomes who are not disqualified by other asset ownership[4] are more suited. This programme was the first to address housing size on a sliding scale of income and household size[5] (KNHC, 2007), and utilizes both the acquisition of existing urban properties as well as new-build housing.

Hybridity in Housing

Contemporary public rental housing is defined as housing built for lower-income people, provided by public or private sectors utilizing public or NHF subsidy, built on publicly developed land, sized less than 85 m^2, and lasting more than 5 years as rental housing (KMLTMA, 2011a; KMOCT, 2004). New programmes, more importantly, have constituted the extension of 'social' housing provision in more or less de-commodified forms for households unable to find adequate housing in the market (Ronald & Lee, 2012). Each new approach has widened eligibility to social housing in response to social, political, and demographic changes. Meanwhile, the scale of public housing has continued to grow: from 6.6 per cent of all housing in 2000 to 9.5 per cent by 2010 (KMLTMA, 2011a; KMOCT, 2002). Social housing provision has not advanced alone, however, and should be considered alongside the rise of civil movements, democratization, and the rolling out of welfare legislation in recent decades.[6]

Table 1 provides a breakdown of KNHC/LH public rental housing programmes concerning production. It illustrates the composite nature of Korean social rental housing, with one residual part (PPRH) serving the interests of the very poor or 'vulnerable' households, and another part made up of two types of public housing that serve low- to middle-income urban workers: LT-PRH for the middle sector of low-income tenants and NPRH (with housing varying in size and price according to economic capacity and household size) for the slightly better off.

Non-permanent or 'fixed tenure' rental housing is a hybrid sector in that tenure is neither fixed for tenant nor dwelling. It is neither social rental housing, nor social home ownership as understood in European terms. It is not yet certain what is to happen once the tenure prescriptions of LT-PRH and NPRH expire. A steady flow from public rental to private ownership of 5-year rental housing is already well established. This tenure explicitly supports home ownership and is not affordable to low-income people.

The ultimate effect of fixed public housing rental tenures is that while provision has increased, it will atrophy with time. This also illustrates a bifurcation in the purpose of contemporary public housing at different times in the housing cycle. At the time of production, all forms of public rental housing perform a social housing role. Meanwhile, toward the other end of the cycle, LT-PRH and NPRH become increasingly commodified. This demonstrates the complexity of public interventions that have become increasingly

Table 1. KNHC/LH public rental housing programmes

Category	Permanent Public Rental Housing (PPRH)	Long-term Public Rental Housing (LT-PRH)	National Public Rental Housing (NPRH)
Planned units (2009–2018)	100 000	200 000	400 000
Realized units (as of 2010)	193 710	913 350	718 493[a]
Tenancy	No limit	50 years[b]	30 years
Unit size (m^2)	23–40	Less than 60 or 60–79	36–52 (Type I); 53–62 (Type II); 63–79 (Type III)
Eligibility	Recipients of Basic Livelihood Security, Persons with distinguished service to the nation	People being relocated from urban redevelopment, Housing Subscription saving account holders, Persons with distinguished service to the nation	Low or moderate-income households with no homeownership (income deciles 1–2 for below 59 m^2; deciles 3–4 for 79 m^2 or larger)
Rent	30 per cent of market rent	80 per cent of market rent	50–80 per cent of market rent
Management	KoHom[c]	KoHom	KNHC/LH, KoHom

[a] including units purchased from privately owned units;
[b] 50-year public rental housing, as a substitute for PPRH, and sized less than 50 m^2, was supplied from 1992 to 1999, and after 2008, the tenancy of newly built PRH was reduced to 10-years and sized between 60 and less than 85 m^2;
[c] KoHom stands for Korea Housing Management Company; KNHC for Korea National Housing Corporation, which was reformed as LH (Korea Land and Housing Corporation) on Oct. 1, 2009.

sophisticated to reflect and target the needs of households of different income levels and household size.

In terms of organizational hybridity, Korean public housing combines features of third sector, state and commercial organizations identified by Brandsen *et al.* (2006). Nonetheless, there is some departure in the nature of hybridity in that social housing remains coordinated by state-owned resources like the KNHC and NHF. Instead of social housing providers becoming more like private independent ones, existing institutions have adapted and enlarged their operations to accommodate more nonmarket housing options, albeit in tenures that often mitigate longer-term de-commodification. Indeed, KNHC was merged with the Korea Land Corporation in 2009, creating the LH, to facilitate an even more centralized and powerful agency.

While this section addressed the impact of public agencies in the expansion, diversification and hybridization of social housing, the next considers how the LH has interacted with private (third sector) and public agencies in housing management and social service delivery.

The Management of Public Housing

The KNHC (or LH) has taken the lead in planning, developing and managing South Korea's burgeoning public housing sector, balancing competing tasks and pressures in each subsector. While it was previously more concerned with supporting private housing supply, in the last two decades it has become embedded in social partnerships with government, civil and voluntary agencies and focused on serving resident's needs and fulfilling welfare functions. On the other hand, it has also had to reconcile affordable housing provision with market contexts. Since the Global Financial Crisis in 2008, pressures to more effectively fund social housing construction and management activities have increased, at the same time the LH has been required to support stability in the market sector.

With the expansion of direct housing provision for socially and economically disadvantaged people, social services have been made increasingly available. PPRH development has legally required the provision of on-site social services since 1989. Currently, most social service centres attached to PPRH support family welfare, community activities, in-home services for vulnerable residents (e.g., meal delivery, medical services, housekeeping, short-term care, etc.), education, job training and counseling. There were 423 social welfare centres registered as of 2010 (KASWC, 2011). Centres are normally operated by either public sector (local governments) or non-profit sector (e.g., religious or corporate organizations, charitable foundations, and voluntary groups). Most services are provided for free or a nominal charge and have been shaped by a raft of legislation developed around the rights and needs of specific users such as the elderly, the disabled, children, and women[7] (KMHW, 2010).

Corresponding to its growing responsibilities for public housing, the KNHC formed a subsidiary company to manage its housing estates, and invested 5-billion KRW[8] in the Korea Housing Management Company (KoHom) in 1998. This organization now manages 250 000 units across more than 300 public estates nationwide (KoHom, 2011). This includes all PPRH and LT-PRH estates, although the latter does not have the same on-site social service requirements for tenants. KoHom has four main tasks: housing management, housing welfare (community activities), upkeep, and renovation. A particular management concern is cost reduction as monthly rental income is very low. While KoHom is the primary institution responsible, it cooperates with partners including 'tenant councils'. The 2000 revision of the Rental Housing Act has meant rental housing estates with 20-plus units can establish such councils giving public tenants some say on the management provided by public landlords.

While KoHom and LH see to housing estate management, related welfare service organizations have emerged, such as Civic Service Centres, Citizens for Decent Housing, Housing Welfare Centres, and Happy House Centres, funded by a mix of public agencies and nonprofit organizations (NPOs). Civic Service Centres are neighborhood-based public offices established by local governments and provide a variety of civic services to support the diverse welfare needs of local people (Nam *et al.*, 2010). Offices are extended to deal with housing related services, including screening and reviewing of applications for pubic rental housing, housing allowance, home repair, *Chonsei* loans, etc. Citizens for Decent Housing, meanwhile, is an NPO (founded in 2001) involved in public policy, community activities, research, and national networks of different interest groups. Its primary concerns are the rights and needs of low-income households as well as building communities in public rental housing estates.

Housing Welfare Centres were established in 2007 and primarily funded by Community Chest of Korea[9] (Nam *et al.*, 2010). On a very limited national annual budget of around 100-million KRW, local centres support community networks and welfare services for low-income and vulnerable households and advocate welfare-related housing rights. The emergence of these centres is attributed to the growing recognition of the role of non-governmental organizations (NGOs) and NPOs in housing policy and welfare support.[10] A similar nonprofit organization, the Korea Housing Welfare Association, was established in 2009 to serve home repair and home improvement for low-income people. There are about 80 entities within this umbrella association (in 2011) that depend on donations from various public and private agencies. These two nonprofit movements in housing have facilitated grass-root community-based activities, filling the gaps in housing service delivery left by public and private sectors. The centres work closely with local governments and public housing agencies (like LH and KoHom), among other private voluntary organizations. Arguably, characteristic of social rental housing elaboration in Korea, then, has been a broader institutional engagement in social welfare with public institutions providing hardware in the form of housing, subsidies, and base services, and privately funded third sector groups enhancing the service levels for vulnerable groups.

The concept of housing welfare has gained in significance subsequent to the 'adequate housing criteria' set out in 2000. Happy House Centres were later implemented as part of the New Housing Movement initiated by the Korea Presidential Commission on Architecture Policy (2009). In 2010, the pilot project established neighborhood centres in three major cities (Seoul, Daegu, and Jeonju). Each centre provides housing-related services[11] free or at nominal cost for lower-income households living in single-family housing units in medium-density neighborhoods. This objective is to promote better quality living environments where single-family homes are highly concentrated. The LH and local governments provide staff and office space for each centre, although operational expenses are funded by the state directly. It is expected that local governments elsewhere will implement their own Happy House schemes subsequent to the success of the pilot.

In contrast to increasing emphasis on housing welfare at the very low-income end of public housing, NPRH has required different interventions. The scale and speed of implementation of NPRH has been remarkable: of the total stock of public rental housing

(1 399 227 units), 681 039 NPRH units were produced between 1998 and 2010 (KMLTMA, 2011a). The LH has become directly involved in the management of many NPRH estates with KoHom responsible for the rest (HURI, 2006; KMLTMA, 2010b). On the one hand, management activities are largely technical (for example, cleaning and security, which are often outsourced). On the other hand, for NPRH, as well as short-term public rental housing orientated toward home ownership, LH and KoHom operate along more commercial lines and also have to answer to more active tenant councils. These are involved in making decisions on bylaws, budgets, long-term planning, daily operations, routine maintenance, and so on. While management offices in housing eventually intended for-sale often implement decisions made by tenant councils, they work on behalf of the public landlords and actively supervise tenants: eligibility screening, rent collecting, tenant training, etc.

Changing Demands on Public Housing Providers

Changes in both provision and management illustrate how public housing has become increasingly 'social' involving a broad elaboration of housing welfare services for low-income households. This has necessitated greater funding and involvement from local governments, NGOs, NPOs, LHA, and in particular the LH. Housing market conditions have also been radically transformed during this period, and price-income ratios in the nation increased from 4.6 in 1997 to 7.7 in 2009 (Kookmin Bank, 2003, 2009). Thus, while the advance of public rental housing has protected more and more poor households, there have been growing pressures to support better off working households for whom conditions have been undermined by volatile market increases. In response, the Moo-hyun Roh administration extended supply and qualification for public housing in the mid-2000s. It also attempted to suppress speculation in the private housing market (Yu & Lee, 2010). These interventions did little to settle housing market volatility (Ha, 2010), a point which was used politically in supporting the transfer from the Roh to the Myung-bak Lee government in 2008.

In the latest version of public housing, the 'Bogeumjari' plan (KMLTMA, 2012), local and national public housing agencies are expected to utilize government or NHF finance to further extend social housing, but in both rented and owner-occupied forms. The *Bogeumjari* program expects to produce 2-million housing units by 2018. Figure 2 sets out how the central government intends to support housing for each income category by the end of the program. This program is more inclusive, consumer-oriented, demand-based, and user-valued in that more income groups are embraced, unit sizes vary with household type and income, rental periods and production mode[12] are diverse, and both public and private providers are various in origin. The public sector will supply 1.5-million units of which more than half will be for rent: 200 000 units for 10-year pubic rental housing, 100 000 units for long-term public *Chonsei* of 10–20-year tenancies, 400 000 units for NPRH, and 100 000 units for PPRH. The private sector will participate in the production of another half-million medium- to large-sized units of 'affordable' owner-occupied housing.

The shift to subsidized home ownership reflects both a decline in private sector production as well as continued affordability problems among middle-income earners. According to KMLTMA (2011a), around 398 000 rental households below income decile 5 are now unable to buy housing without public support. A nascent housing allowance/voucher scheme is thus also being developed for low-income families, while

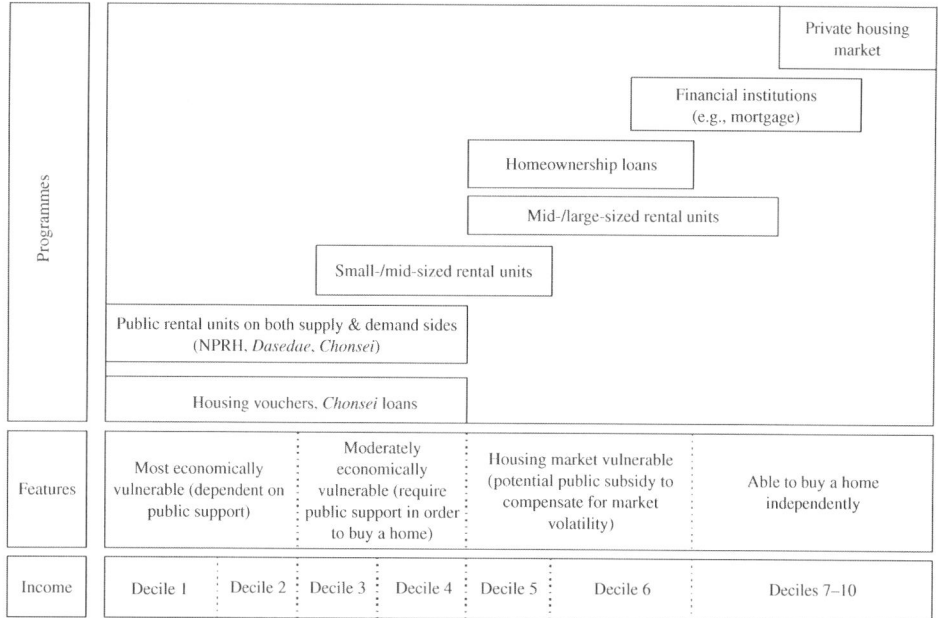

Figure 2. Blueprint of housing welfare in 2017. *Note: Dasedae* is one type of multifamily housing made up of four-storey buildings, built by private developers with subsidized loans from public agencies. *Source*: KMLTMA (2010a).

public *Chonsei* loan schemes and low interest mortgages have also been made available for moderate-income households. The *Bogeumjari* plan demonstrates both comprehensiveness and complexity in how different providers will produce, allocate, and manage distinctive housing types and tenures for different income groups. It also represents another level of hybridization in the organization of social housing, which is becoming split four ways: the most vulnerable in PPRH, low-income renters in 10-year PRH; moderate-income quasi-temporary renters in NPRH and subsidized *Chonsei*, and a new tranche of social owner-occupiers and subsidized borrowers.

Since the Global Financial Crisis in 2008, however, there have been the unanticipated challenges of housing oversupply. The housing market has become sluggish with, as of 2010, as many as 165 000 unsold units nationally (HERI, 2010). In this context, pressure has been placed by the state on the LH to buy up more land and property to support the market. At the same time, it has been increasingly unable to unload redeveloped land or sell properties (Korea Times, 2010). In 2011, the LH assets amounted to 148-trillion KRW making it the second largest industrial corporation in the country. In all, the LH carries out land acquisition and urban/land development and regeneration, as well as housing production, management, and renovation on behalf of the central government. The initial capital finance from the government in 2010 was 22-trillion KRW. However, the LH debt in 2010 was 118-trillion KRW, while the proportion of liabilities to LH total assets more than quintupled. The capacity of the LH has thus been stretched, perhaps to its limits. Figure 3 illustrates the increasing scale of debt the LH has taken on in recent years, which in 2010 represented a sum equivalent to 10 per cent of GDP.

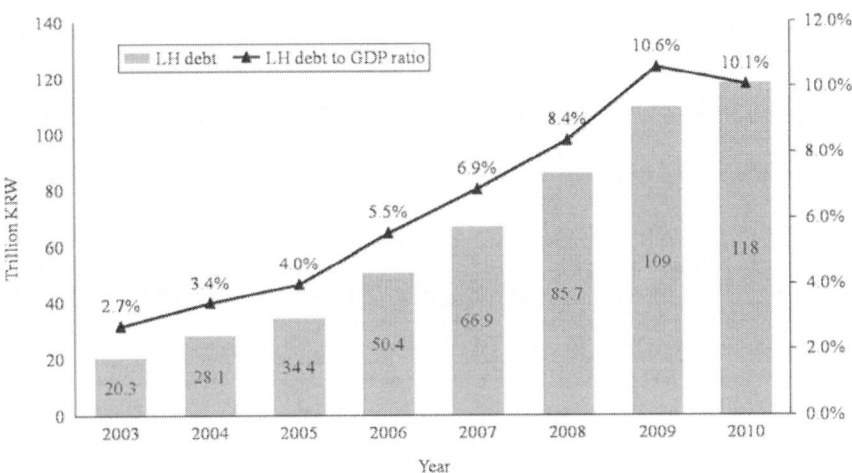

Figure 3. LH's debt and debt-to-GDP ratio. *Source*: LH unpublished data.

LHA have also been stretched by governments pushing them to support urban redevelopment, home building, and the subsidization of affordable rents. Meanwhile, the economic resources needed to implement such projects have been relatively limited. In Korea, 16 of the 131 local public agencies are housing agencies. It has been reported that total local public agency debt reached 42.6-trillion KRW in 2009, with LHA accounting for about 82 per cent (*Donga Ilbo*, 2010). The average debt ratio of LHA was 2.5 times (347.1 per cent) greater than that of local public agencies (132.8 per cent). The debt of the Seoul Metropolitan Area (including SH, GICO, and IUDC) accounts for more than three quarters (78.7 per cent) of total LHA debt (27.5-trillion KRW). Table 2 provides a breakdown of this debt among LHA, which has been further enhanced in the last few years by the economic downturn and a sluggish real estate industry. There is a growing likelihood that projects led by LH and LHA may be scaled back as a result of continued problems in finance, undermining social housing capacity.[13]

Discussion

The 'modernization' of South Korean public housing demonstrates a clear departure from European experiences. Korea achieved a high level of economic development[14] without elaborating its welfare state. In light of declining housing affordability and intensified political contestation, housing and social policy have, however, become more central to political debates. Meanwhile, government interventions have shifted, in the context of the post-Asian Financial Crisis era, from serving rapid growth to sustaining the impetus of growth while managing the volatility of and socioeconomic inequalities generated by the market. These transformations have provided a particular framework for the hybridization of public housing provision, requiring both expansion and diversification.

The LH as provider and manager has become—along with the elaboration of different sectors in response to changing sociopolitical and market pressures at different times— clearly more hybridized. In contrast to European hybridization, however, this is not a result of state withdrawal from welfare provision and pressures on social landlords to

Table 2. Debt ratio of public housing agencies (in 2009)

Public housing agencies	Year founded	Capital (billion KRW)	Debt (billion KRW)	Debt ratio (percent)[a]
LH (state-owned housing agency)	2009	21 000	109 000	519
Local housing agencies (LHA)				
Seoul Housing Corporation (SH)	1989	3233.3	16 345.5	505.5
Busan Metropolitan Corporation (BMC)	1991	763.1	2167.1	284.0
Daegu Urban Development Corporation (DUCO)	1988	372.5	708.9	190.3
Incheon Urban Development Corporation (IUDC)	2003	1850.7	4460.9	241.0
Gwangju Metropolitan City Corporation (GMCC)	1993	241.0	618.3	256.5
Daejeon City Corporation (DCCO)	1993	208.4	416.2	199.7
Ulsan Metropolitan City Authority (UMCA)	2007	85.8	296.9	346.0
Gyeonggi Urban Innovation Corporation (GICO)	1997	1708.7	6715.9	393.0
Gangwondo Development Corporation (GDCO)	1997	335.3	1048.8	312.8
Chungbuk Development Corporation (CBDC)	2006	133.4	251.1	188.2
Chungnam Development Corporation (CNDC)	2006	211.2	333.8	158.1
Jeonbuk Development Corporation (JBDC)	1998	96.7	341.8	353.6
Jeonnam Development Corporation (JNDC)	2004	495.7	418.8	84.5
GyeongSangBuk-Do Development Corporation (GBDC)	1997	106.7	123.5	115.7
Gyeongsangnam-Do Development Corporation (GNDC)	1997	154.1	679.7	441.0
Jeju Special Self-governing Province Development Corporation (JPDC)	1995	83.0	54.8	66.0
Total (LHA)		10 079.6	34 982	

[a] Debt ratio is calculated by dividing total liabilities of each agency by its total assets.
Sources: A Report of the Korea Ministry of Public Administration and Security submitted to National Assembly (*Donga Ilbo*, 2010).

become more multifaceted social enterprises, but part of the expansion of public provision and the role of public agencies. Indeed, the proportion of social housing, for-sale or rent, is expected to exceed 12 per cent of all housing by 2018, up from 7.5 per cent in 2000 (KMLTMA, 2008; KMOCT, 2002). The LH has not streamlined or come to resemble an entrepreneurial private agency. Rather it has become bulkier and even more dependent on state finance. Moreover, in light of the economic success of developmentalism, government agencies have been considered competent providers and managers of housing, justifying their expansion and reliance on public coordination. Indeed, the trend has been toward more centralization and intensified direct public funding for social housing.

Although Korean public housing has become more social, modes of delivery have been hybridized in terms of semi-market forms of provision along with market supportive interventions. In recent decades, a social housing sector has been not only embedded, but also split between, firstly, permanent public rental; secondly, long- and short-term public rental; and, thirdly, public owner-occupied housing. Essentially, each of these subsectors sits in relation to the next on a de-commodified/commodified continuum with each serving different social or income constituencies with different tenure conditions according to how each group has been considered market excluded. In this sense, Korean public housing represents both an organizational and a modal hybridization. In the latest *Bogeumjari* plan, more than half of rental households, based on income criteria, qualify for some form of public housing, reflecting features of a *universal/generalist* approach to social housing provision. Meanwhile, PPRH has been revived to target the most vulnerable, reflecting *residual* tendencies.

By providing housing goods and services for non-economically productive sectors of society, recent approaches have marked a departure from past developmental models of public housing. While Korea provides an extreme case, it reflects similar trends in the region toward social housing measures in context of market problems. Housing affordability has become a core public concern in China, Hong Kong, Singapore, and Taiwan, where social housing measures have been revamped since the mid-2000s (Ronald & Chiu, 2010).

The Korean approach strongly reflects a path shaped by the formation of large interventionist entities, such as the KNHC, based upon developmentalist imperatives, with adaptation reacting to shifting pressures in the 1990s and 2000s to improve housing conditions for the increasing numbers of households left behind by the market. Public housing has not conflicted with the market, but sought to either stimulate it or, as is increasingly the case, prop it up by accommodating those excluded; facilitate market entry through temporary rental tenures and home ownership schemes, and, more recently, buying up existing property and land to support market prices. After more than three decades of real estate-driven capital accumulation, diminishing affordable housing conditions has pressed the state to redefine the role of the LH and other public housing agencies. This has, nonetheless, stretched institutional reserves of housing associations.

Conclusion

Oxley (2000) identifies three criteria that have effectively defined social housing in Western contexts: housing owned by a social landlord, housing provided at below market price, and housing allocated by need and not ability to pay. While Western concepts seem relevant to the Korean case, they are neither equivalent nor transferable. What is shared with contemporary European social housing organizations is the triangular orientation of housing agencies (Brandsen *et al.*, 2006), with parts facing the market and community, but with the state still playing an important role. Korean housing operations have, however, neither followed a path involving early expansions of social rental housing and welfare state services, nor new public management trends focused on competition and private sector performance culture. Indeed, features of new public management are not immediately apparent as, arguably, developmentalism has protected public sector privileges. Nonetheless, transformations in civil society, political life, and socioeconomic conditions are dramatically changing expectations of public and social housing organizations.

Despite the scale of transformation in South Korea, there has been little comparative examination of differences across social housing sectors within East Asia or with other developed societies. Korea throws up important questions for understanding processes of neoliberalization and the ways different societies are responding to shifting socio-economic and demographic conditions, as well as the implications of changing market and welfare demands for housing provision.

East Asian housing systems and policy models constitute an important challenge to European-derived understandings of what drives social housing development and how housing policy reacts to socioeconomic and political developments. Hybridity is an important concept that applies to different contexts in ways that have until now been inadequately conceptualized. While Korea demonstrates quite peculiar features of hybridization, Wang & Murie (2011) also illustrate in China that despite intensive housing marketization, nonmarket housing options have also been advanced by the state, at quite a different stage in economic development, resulting in another particular 'hybrid system' that reflects transitional legacies and indigenous processes of urbanization and demographic change. The further examination and comparison of East Asian contexts thus presents an opportunity to challenge understandings of social and public housing that have been derived and distilled with a largely Eurocentric bias.

Notes

[1] The first local public housing agency, founded in 1988, was Daegu Urban Development Corporation (DUCO), while the largest is the Seoul Housing Corporation established by Seoul Metropolitan Government in 1989.

[2] *Chonsei* renters pay, in principle, no monthly rent, but rather an up-front sum of up to 70 per cent of the value of the property for a fixed period of tenancy, typically 2 years. Large deposits provide landlords with capital for informal banking activities and housing investment, with profits serving in lieu of rental income. At the end of the lease, the deposit is returned to the tenants.

[3] In 2008, PPRH production was resumed as part of the state-led housing program, 'Bogeumjari'. The current plan is expected to supply another 100 000 new units between 2009 and 2018.

[4] Qualification requires households to have an income 10 per cent below the urban average and no assets valued more than 50-million KRW (or owning a car worth more than 22-million KRW).

[5] The smallest units (under 53 m^2) are designated for households with incomes 50 per cent or below the monthly average of working households, with larger units allocated to moderate-income families.

[6] Civil Movement in 1987, Child Care Act of 1991, the National Basic Livelihood Security Act of 1999, Act on Low Fertility and Aging Society in 2005, Affirmative Action for Women in 2006, and Disability Discrimination Act of 2007.

[7] The Social Security Act of 1995; the Social Welfare Service Act of 1997; the National Assistance Act of 1961; the National Basic Livelihood Security Act of 1999; the Elderly Welfare Act of 1961; the Act on Low Fertility and Aging Society, 2005; the Handicapped Persons' Welfare Act of 1981; the Disabled Persons' Welfare Act of 1989; the Disability Discrimination Act of 2007; the Child Welfare Act of 1961; the Early Childhood Education Act of 1982; the Child Care Act of 1991; the Women's Rights Act of 1995; the Affirmative Action for Women of 2006.

[8] Approximately 1150 KRW to 1 USD

[9] A Korean NGO founded in 1998 and a member of United Way International, the nation's largest charity that is committed to empowering people through their communities.

[10] The National Society of Housing Welfare Centres was formed in 2010.

[11] Counseling, home safety checks, emergency home repair services, fitting energy-saving lighting, insulation, and landscaping.

[12] The *Bogeumjari* program also involves the market acquisition of existing units in addition to construction-based supply.

[13] Financial problems largely stem from excessive financing of post-1998 housing and urban development projects. In an aggressive effort to reduce the financial burden, the LH has recently begun canceling new urban developments and delaying planned housing projects.

[14] Achieving OECD status in 1996.

References

Brandsen, T., Farnell, R. & Cardoso Ribeiro, T. (2006) *Housing Association Diversification in Europe: Profiles, Portfolios and Strategies* (Coventry: The Rex Group).

Castells, M., Goh, L. & Kwok, R. Y. -W. (1990) *The Shek kip Mei Syndrome: Economic Development and Public Housing in Hong Kong and Singapore* (London: Pion).

Chiu, R. L. H. (2008) Government intervention in housing: convergence and divergence of the Asian Dragons, *Urban Policy and Research*, 26(3), pp. 249–269.

Chua, B. H. (1997) *Political Legitimacy and Housing: Stakeholding in Singapore* (London: Routledge).

Cowans, J. & Maclennan, D. (2008) *Visions for Social Housing: International Perspectives* (London: Smith Institute).

Czischke, D. (2009) Managing social rental housing in the EU: A comparative study, *European Journal of Housing Policy*, 8(2), pp. 121–151.

Czischke, D. K., Gruis, V. H. & Mullins, D. W. (2010) Conceptualizing social enterprise in housing organisations. Paper presented at the ENHR Conference on Urban Dynamics and Housing Change, Istanbul. 4–7 July 2010.

Doling, J. (1999) Housing policies and the little tigers: how do they compare with the other industrialised countries, *Housing Studies*, 14(2), pp. 229–250.

Donga Ilbo (2010, April 4) Local public housing companies and their growing debt. Available at http://news.donga.com/3/all/20101003/31584590/1

Ghekiére, L. (2007) *Le développement du logement social dans l'Union Européenne. Quand l'intérêt général rencontre l'intérêt communautaire* (CECODHAS-USH-Dexia) (Paris: Dexia Editions).

Gill, I. & Kharas, H. (2009) Gravity and friction in growing East Asia, *Oxford Review of Economic Policy*, 25(2), pp. 190–204.

Ha, S. K. (2010) Housing crises and policy transformations in South Korea, *International Journal of Housing Policy*, 10(3), pp. 255–272.

HERI (Hyundai Economic Research Institute) (2010) *Homeowners Using Bank Loans May Fall into the Deadly Spiral of Debt Deflation* (Seoul: HERI).

HURI (Housing and Urban Research Institute) (2006) *Whitepaper on National Public Rental Housing* (Seongnam: Korea National Housing Corporation).

KASWC (Korea Association of Social Welfare Centers) (2011) *A Brief Report of Social Welfare Centers in Korea* (Seoul: KASWC).

Kemeny, J. (1995) *From Public Housing to the Social Market: Rental Policy Strategies in Comparative Perspective* (London: Routledge).

KMHW (Korea Ministry of Health and Welfare) (2010) *2009 Whitepaper on Health and Welfare* (Gwacheon: KMHW).

KMLTMA (Korea Ministry of Land, Transport and Maritime Affairs) (2008) *A Manual of 2008 Rental Housing* (Gwacheon: KMLTMA).

KMLTMA (2010a) *2010 Housing Yearbook* (Gwacheon: KMLTMA).

KMLTMA (2010b) *2010 Housing Plan* (Gwacheon: KMLTMA).

KMLTMA (2011a) *2011 Housing Yearbook* (Gwacheon: KMLTMA).

KMLTMA (2011b) *A Manual of 2011 National Housing Fund* (Gwacheon: KMLTMA).

KMLTMA (2012) Bogeumjari overview. Available at http://www.newplus.go.kr/newplus_theme/portal/newplus/newplusIndex.page

KMOCT (Korea Ministry of Construction and Transportation) (2002) *2002 Housing Yearbook* (Gwacheon: KMOCT).

KMOCT (2004) *2004 Housing Yearbook* (Gwacheon: KMOCT).

KNHC (Korea National Housing Corporation) (2005) *Yearbook of Housing and Urban statistics* (Seongnam: KNHC).

KNHC (2007) *A Handbook for Rental Housing and Housing Welfare Program* (Seongnam: KNHC).

KoHom (Korea Housing Management Company) (2011) Company profile. Available at http://kohom.co.kr/intro/intro01.jsp

Kookmin Bank (2003) *2003 Home Financing and its Demand Report* (Seoul: Kookmin Bank).

Kookmin Bank (2009) *2009 Home Financing and its Demand Report* (Seoul: Kookmin Bank).

Korea Presidential Commission on Architecture Policy (2009) *Overview of Presidential Commission on Architecture Policy* (Seoul: Korea Presidential Commission on Architecture Policy).

Korea Times (2010) LH's unsold poperties reach $20 billion. Available at http://www.koreatimes.co.kr/www/news/biz/2010/10/123_70452.html

Kwon, H. J. (2009) The reform of the developmental welfare state in East Asia, *International Journal of Social Welfare*, 18(s1), pp. S12–21.

Lee, H. & Hong, H. (2007) An examination of housing policy for low-income households in Korea. Paper presented at the Asia Pacific Network for Housing Research (APNHR) Conference—Transformation in Housing, Urban Life, and Public Policy, Seoul National University, South Korea, August.

Lim, S. H. (2005) *A Half Century of Housing Policy* (Seoul: Gimundang).

Malpass, P. & Victory, C. (2010) The modernisation of social housing in England, *International Journal of Housing Policy*, 10(1), pp. 3–18.

Nam, W., Choi, E. & Cho, K. (2010) *A Study on Designing Providing System of Community-Based Housing Support Services* (Daejeon: Land and Housing Research Institute, Korea Land and Housing Corporation).

Oxley, M. (2000) *The Future of Social Housing: Learning from Europe* (London: IPPR).

Park, B. G. (1998) Where do tigers sleep at night? The state's role in housing policy in South Korea and Singapore, *Economic Geography*, 14, pp. 229–250.

Park, S. Y. (2007) The state of housing policy in Korea, in: R. Groves, A. Murie & C. Watson (Eds) *Housing and the New Welfare State: Perspectives from East Asia and Europe*, pp. 75–100 (Aldershot: Ashgate).

Peng, I. (2004) Postindustrial pressures, political regime shifts and social policy reforms in Japan and South Korea, *Journal of East Asian Studies*, 4(3), pp. 389–425.

Priemus, H. & Dieleman, F. (2002) Social housing policy in the European Union: past, present and perspectives, *Urban Studies*, 39(2), pp. 191–200.

Priemus, H. & Gruis, V. (2011) Social housing and illegal state aid: the agreement between European commission and Dutch government, *International Journal of Housing Policy*, 11(1), pp. 89–104.

Rhodes, M. L. & Mullins, D. (2009) Market concepts, coordination mechanisms and new actors in social housing, *European Journal of Housing Policy*, 8(2), pp. 107–119.

Ronald, R. (2007) Comparing homeowner societies: can we construct an East–West model? *Housing Studies*, 22(4), pp. 473–493.

Ronald, R. & Chiu, R. H. L. (2010) Changing housing policy landscapes in Asia Pacific, *International Journal of Housing Policy*, 10(3), pp. 223–231.

Ronald, R. & Jin, M. -Y. (2010) Homeownership in South Korea: examining sector underdevelopment, *Urban Studies*, 47(11), pp. 2367–2388.

Ronald, R. & Lee, H. (2012) Housing policy socialisation and de-commodification in South Korea, *Journal of Housing and the Built Environment*. DOI: 10.1007/s10901-011-9257-2.

STFSAP (Special Task Force for State Affairs Press) (2007) *Forty Years of Real Estate in Korea* (Seoul: Hans Media).

Wade, R. (1990) *Governing the Market: Economic Theory and the Role of Government in East Asian Industrialisation* (Princeton, NJ: Princeton University Press).

Walker, B. (2001) The changing management of social housing: the impact of externalisation and managerialisation, *Housing Studies*, 15(2), pp. 281–299.

Wang, Y. -P. & Murie, A. (2011) The new affordable and social housing provision system in China: implications for comparative housing studies, *International Journal of Housing Policy*, 10(3), pp. 237–254.

Yu, H. -J. & Lee, S. (2010) Government housing policies and housing market instability in Korea, *Habitat International*, 34, pp. 145–153.

Negotiating Tensions: How Do Social Enterprises in the Homelessness Field Balance Social and Commercial Considerations?

SIMON TEASDALE

Third Sector Research Centre, University of Birmingham, Birmingham, UK

ABSTRACT *Social enterprise is presented as a potential policy solution to homelessness, particularly as regards the employment of homeless people. This policy focus relies on an assumption that social and commercial goals can be successfully combined. This implies that by pursuing profit-maximizing behaviour social enterprises can also maximize social benefits. However, this paper shows that social enterprises are hybrid organizations facing a trade-off between social and commercial considerations. The paper identifies strategies used by work integration social enterprises in the homelessness field to balance mission-related goals with financial sustainability. The six case study organizations drew upon a hybrid range of economic resources transferred from other sectors of the economy. This enabled them to compete with private sector organizations, by effectively transferring the additional cost of employing homeless people from the social enterprise to consumers, government, philanthropic donors, and other organizations providing social support to homeless people.*

Introduction

Social enterprises, broadly defined as organizations trading in the marketplace to achieve a social purpose (DTI, 2002), have achieved widespread recognition over the last decade. In England, the notion popularized in academic literature that social enterprises successfully combine social and economic goals led to them being given a prominent role in a range of policy areas (Teasdale, 2011). This has been particularly apparent in the homelessness field where a policy discourse presents social enterprise as a way to help homeless people access secure employment and so escape social exclusion.

Simultaneously, government has presented social enterprise as a sustainable funding mechanism for voluntary and community organizations, so lessening their dependence on philanthropy (and presumably the public purse). The implication is that by pursuing profit-maximizing behaviour social enterprises can address homelessness while generating a surplus from their trading activities that can be reinvested in the business or used for other mission-related purposes.

From this perspective, social enterprise appeared a win-win game. A range of 'how to' texts emanated from business schools instructing voluntary organizations how to develop new revenue from trading activities (Dey & Steyaert, 2010). The application of commercial logics to social problems that had proved beyond bureaucratic governments and well meaning but inefficient charities (Dees, 2007) was popularized as a way to 'change the world' (Bornstein, 2001). However, some commentators argued that social enterprise was not so much 'new' as a re-packaging of existing phenomena. Mullins & Riseborough (2000) note that Housing Associations were trading for a social purpose before the term social enterprise entered the policy landscape. Defourny & Nyssens (2010) argue that social enterprises in Europe have emerged from earlier co-operative models that also sought to trade for a social purpose. Here, the assumption is that social enterprises are more akin to hybrid organizations seeking to blend the operational priorities of the third sector (distinctive social mission) with those of the private sector (market forces) (see Billis, 2010). But can these priorities be successfully combined as suggested by the policy focus in England, or must organizations prioritize one over the other?

Certainly, past experience would suggest social enterprises face inherent tensions between the competing logics of market and third sector. The failure of co-operatives to play more than a marginal role in the economy has been attributed to a degeneration thesis that proposes that over time the democracy inherent in early stage co-ops becomes overwhelmed by the dominant market logic and sees them behaving like private firms (Cornforth et al., 1988). These isomorphic tendencies have also been identified in 'new' forms of social enterprise. While organizations may be initiated to meet social objectives, the realities of adapting to a capitalist system favouring cost reduction leads many to become indistinguishable from mainstream businesses (Amin et al., 2002). Isomorphism may be particularly apparent in work integration social enterprises (Aiken, 2006; Gardin, 2006) seeking to 'employ or train marginalized people within an enterprise that has social dimensions, and that trades in the market' (Spear & Bidet, 2005, p. 197).

This paper identifies implicit and explicit strategies used by work integration social enterprises in the homelessness field to balance the contradictory social and commercial logics of employing homeless people and competing in the private marketplace. The paper begins with an overview of the social enterprise policy background, paying particular attention to the role of social enterprise in providing employment to homeless people. This is followed by the introduction of the analytic framework used to understand the behaviour of social enterprises. As a conceptual starting point, it is assumed that balancing social and commercial considerations is a zero-sum game. Organizations must forego a social return to maximize financial revenue (and vice versa). The 'Methods' section outlines the approach taken to data collection and analysis in this qualitative exploratory study. In the 'Findings' section, strategies are identified that enable social enterprises to re-negotiate the boundaries implied by the social–commercial tension. These involve drawing upon additional resources that private firms (or charities) cannot normally access. The main argument and contribution of this paper, developed fully in the concluding section, is that social enterprises can compete with private sector organizations, by drawing upon hybrid resource mixes that effectively transfer additional costs to other resource holders. Thus, social enterprises are beneficiaries of resource transfers from donors and investors, government, parent charities, volunteers, consumers, and other organizations providing social support to homeless people. These resource transfers provide social enterprises with a competitive advantage that can offset the extra costs associated with employing homeless people whose productivity may be low.

Policy Background

Before the election of a Conservative-led coalition government in 2010, England had perhaps the most developed institutional support structure for social enterprise in the world (Nicholls, 2010). Under the 1997–2010 Labour government, considerable coercive pressure was placed on voluntary organizations to develop as social enterprises to receive public funds (Carmel & Harlock, 2008). An underlying policy assumption was that social issues and commercial opportunities could be successfully combined to produce 'business solutions to achieve public good' (DTI, 2002, p. 7). In relation to homelessness, an early indication of the possible role for social enterprise came in the then Office of the Deputy Prime Minister's Homeless Statistics report for June 2003 that saw social enterprise as a first step to employment for many homeless people:

> Social enterprises have a distinct and valuable role in helping create a strong, sustainable and socially inclusive economy. For many homeless people engaging with a social enterprise is a first step towards mainstream employment. (ODPM, 2003, p. 9)

In 2005, the briefing accompanying the launch of the 'Hostels Capital Improvement Programme' suggested that hostels and day centres for homeless people might include spaces for social enterprises to encourage self-employment, develop self-confidence and esteem, and offer routes into employment (ODPM, 2005). The Hostels Capital Improvement Programme was to spend £90 million over three years on improving the physical environment occupied by homeless people, and aimed to move them into employment or training and a settled home (CLG, 2006). It was followed in 2008 by 'Places of Change', a three-year £70 million programme, which paid specific attention to the role of social enterprise in moving homeless people into employment (CLG, 2007).

The government role in promoting social enterprise involved attempting to increase the supply of social enterprises offering employment training for homeless people. This continued with the launch of the 2007 SPARK initiative. SPARK involves a 'dragons den' style competition for initiatives in public, private, and voluntary sectors that use social enterprise to tackle homelessness. Announcing government funding for the 2009 'SPARK challenge', the then Homeless Minister Ian Wright (2008) said:

> Tackling homelessness requires new and innovative solutions and this is exactly what SPARK makes possible. I would encourage social enterprises to grasp this great opportunity not only to obtain funding, but also to get expert advice from leading businesses to both grow as an enterprise and be able to help change the lives of many more people.

The 2009 competition offered a total prize fund of £3.4 million, predominately funded by central government. It attracted 139 entrants of which 15 were 'winners'. All of these were involved in work integration for homeless people. It is important to note the strong policy message that employment is the sustainable solution to homelessness, and that work integration social enterprises are primarily placed as a vehicle that can facilitate this solution. This message would seem to have found favour with the new coalition government in the UK, which supported SPARK 2011, and continued wider investment in work integration through the Work Programme.

Social Enterprise

Social enterprise is a label that is used to describe a wide variety of organizational types and activities across the world (Kerlin, 2010). There is no commonly accepted definition of social enterprise, although there is some consensus that it involves organizations that trade in the marketplace to meet their social goals (Peattie & Morley, 2008). These social enterprise organizations can be conceptualized as varying by field of activity (Young, 2010) and by organizational type or model (Alter, 2007). In England, a (deliberately loose) definition drafted by what was the government Department for Trade and Industry in partnership with a range of practitioner organizations has been widely adopted:

> A social enterprise is a business with primarily social objectives, whose surpluses are principally reinvested for that purpose in the business or in the community, rather than being driven by the need to maximise profit for shareholders and owners. (DTI, 2002, p. 8)

The ambiguity around this definition, particularly around the ways in which 'social objectives' and the reinvestment of surpluses in the business, has meant that a wide range of organizations are able to claim to be a social enterprise, or have had the label attributed to them (Lyon *et al.*, 2010; Teasdale, 2011).

Early academic literature presented social enterprise as a way for nonprofits to generate a surplus to be re-invested in their social mission-related activities (Dees, 1998). This social enterprise activity could be *external* to the primary activity of the organizations (Alter, 2007). So, for example, Shelter derives funds through the sale of Christmas cards to subsidize its work with homeless families (Teasdale, 2010a). Alternatively, the trading activity can be *embedded* within the social mission (Alter, 2007). For example, *The Big Issue* derives the majority of revenue through a magazine sold by homeless people. In both cases, it would initially appear that social and economic goals are in alignment. The more Christmas cards sold by Shelter, the more money can be diverted to social activities. Similarly, the more magazines sold by *The Big Issue*, the more homeless people that are able to earn an income. The implicit assumption is that by pursuing profit-maximizing behaviour social enterprises are able to achieve their social goals.

Much of the early academic work sought to identify and explain this 'new' phenomenon of social enterprise. Studies tended to be largely descriptive and drew upon illustrative rather than analytical cases (Short *et al.*, 2009). The assumption that social enterprises successfully combined social and economic goals was largely unquestioned (Dey & Steyaert, 2010). As the field of social enterprise research developed, more critical commentaries began to question this assumption theoretically (Blackburn & Ram, 2006) and empirically (Amin *et al.*, 2002;

Russell & Scott, 2007; Teasdale, 2010b). It became apparent that social enterprises face an inherent tension between social and economic objectives. Returning to the example of *The Big Issue*, although social and economic objectives would appear in close alignment in this embedded social enterprise, if the social objective is to provide homeless people with an income, this would better be achieved by reducing the price of the magazine paid by homeless vendors. However, this would reduce the organization's income. Thus, setting the price of the magazine paid by homeless vendors involves a trade-off between social and commercial objectives.

It is widely accepted that social enterprise lies towards the centre of a spectrum of organizations ranging from the traditional charity to the neoclassical private firm (Dees, 1998). There is, however, some dispute as to where the boundaries lie between social enterprise and charities on the one side, and private businesses on the other. A narrow definition would tend to focus on social enterprises whereby social goals are directly *embedded* into the trading activity, for example, *The Big Issue*. Wider definitions, particularly in the USA have extended from corporate social responsibility (CSR) initiatives by private firms to non-mission-related revenue-raising activities by charities (Lyon *et al.*, 2010). Once the tension between social and economic goals is incorporated into this spectrum (see Figure 1), it is possible to conceptualize the neoclassical conception of the private firm as occupying one pole. Here, the aim is solely to maximize commercial profit to provide a return on investment for owners. There is no duty (or moral responsibility) to provide a social return (Friedman, 1970). Occupying the opposite pole are those charities that (in theory) maximize their social return while foregoing any financial return. Between these two hypothetical extremes are a range of organizations

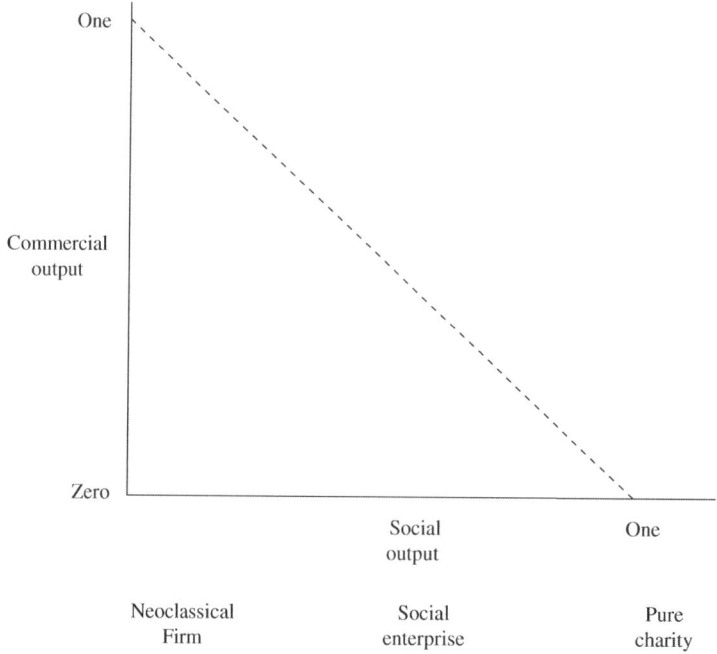

Figure 1. Balancing social and commercial considerations.

from multi-national organizations with a CSR agenda to charities deriving some revenue from, for example, the sale of cakes at a church fair. For the (illustrative) purposes of this paper, it is initially assumed that negotiating social and commercial considerations involves a straight trade-off. Thus, as a starting point, all organizations can be placed at a point along the spectrum. This position equates to a specified social output and a specified commercial output, the sum of which is equal to 1.

The tensions between social and commercial considerations are particularly apparent among embedded social enterprises in the field of work integration. A major study on work integration social enterprises employing or training disadvantaged groups found they faced considerable pressure to cream off the least disadvantaged to reduce costs (Nyssens, 2006). This applies particularly when an organization is paid by government on an 'outcomes' basis to prepare people for placement into the mainstream labour market (Aiken & Bode, 2009). In the field of homelessness, this is because homeless people have varying levels of social and economic needs (Pleace, 1999; Stephens & Fitzpatrick, 2007). Those with the highest support needs are harder to place into employment than those with lower support needs. Hence, a social enterprise paid for each person placed into employment has a 'natural' tendency to cream off the least disadvantaged (Teasdale, 2010a), and effectively move from the social pole towards the economic pole of the spectrum in Figure 1.

Another model of embedded social enterprise prominent in the homelessness field involves organizations directly providing paid (or unpaid) employment to homeless people within a commercial business competing in the open market (Teasdale, 2010a). Disadvantaged people may have a lower commercial value to the organization than labour sourced from the open market. This is because they require additional social support and so are less economically productive (Mozier & Tracey, 2010; Tracey & Jarvis, 2007). Returning to Figure 1, this additional financial cost to the employer can be theorized as equating to additional social output. That is the extra costs of providing social support to homeless people are offset by the higher social return achieved from employing them. However, embedded social enterprises of this type are not usually paid a direct commercial return for achieving this social output. Instead, they have to maximize commercial revenue to provide ongoing social support to homeless employees (Tracey *et al.*, 2010). In the 'efficient' private firm, one way of achieving this would be to reduce employment costs by cutting salaries (Hansmann, 1987). However, in a social enterprise this would reduce the social output of the organization as those same employees are the beneficiaries the organization serves to exist (Tracey & Jarvis, 2007). How then do embedded work integration social enterprises in the homelessness field balance these particular social and commercial considerations, and what strategies do they employ to maximize their aggregate social and commercial return? In particular, is it possible for some social enterprises to move beyond the limits of aggregate social and commercial output implied by the model outlined in Figure 1?

Methods

A qualitative case study approach was adopted to address these questions. The emphasis was on those organizations directly engaging homeless people in the trading activity, whether as trainees or paid employees. There are no reliable estimates of the numbers of social enterprises in the homelessness field (McKenna, 2011), or indeed the wider social enterprise population in the UK. This is in part because the concept of social enterprise is contested with most rigid definitions deriving from the normative views of different commentators (Lyon *et al.*, 2010). Therefore, sampling was purposive, aiming to capture organizations adopting each of these approaches. Organizations were selected from members of a national network of self-identifying social enterprises in the homelessness field. The aim was not to construct a representative sample, but to capture a diverse range of work integration social enterprises. All engaged in trading for some proportion of their income (see Table 1). None distributed profits to external shareholders.

A scoping study (Teasdale, 2010a) identified two broad types of work integration social enterprise activities. Differences centred upon whether the homeless participant was unpaid (the training and work experience model) or paid (the employment provider model). Training and work experience could be provided within a commercial enterprise, a participation-based community where employment was one element of a holistic package aimed at reintegrating homeless people into mainstream society, or as part of a government-funded welfare to Work Programme that paid the social enterprise for each homeless person trained and moved into employment. One organization from each of these sub-types was selected. Three organizations from the employment provider model were selected, making a total of six cases (see Table 1).

Headline descriptive data was collected through the analysis of annual reports, company websites, and financial data. Primary data was collected through semi-structured interviews with key staff at each of the organizations. Pilot interviews highlighted that many interviewees would not recognize at an abstract level any tension between social and economic objectives. Instead, interviews aimed to identify specific examples that illustrated these themes. A total of 18.5 hours of transcripts were stored and manipulated using NVIVO (computer-aided qualitative analysis software). Data analysis was partly deductive in that pre-defined themes around social and economic objectives were imposed upon data collection, and the first stage of analysis. These themes were developed through a more inductive and iterative approach that involved moving back and forth between data, emerging propositions, and existing academic literature. The use of multiple cases allowed for replication to test and refine theories as they emerged (Yin, 2003). The purposive sampling allowed the exploration of difference between cases.

Introducing the Cases

Social Goals

Each case study organization had the employment of homeless people as a social objective. For all but Environment Training, this was a primary social objective as expressed in their organizational mission statement. Environment Training was a social enterprise delivering government contracts around welfare to work and health outcomes. One of these programmes specifically aimed to move homeless people into employment. Environment Training provided accredited training courses to participants and through a consortium aimed to place them into full-time employment in the mainstream labour market.

Other approaches to moving homeless people into employment diverged primarily around whether the organization provided paid employment to participants, or instead offered training and work experience. For example, Garden Furniture Training provided work experience and non-accredited training to homeless people in an enterprise making

Table 1. Introduction to the case study organizations

Case study	Type	Commercial approach	Non-commercial resources	Social goals
Dry Cleaning Works	Employment provider	Provides same day cleaning service to individuals working at city firms: 50–75 per cent commercial income.	Salaries of key staff paid by grant. Free building rental.	To provide homeless people with temporary paid employment as a stepping stone to the mainstream labour market.
Removal Works	Employment provider	Provides removals service to local businesses: 75–100 per cent commercial income.	Salaries of key staff paid by grant. Free building rental.	Permanent employment of formerly homeless people.
Street Cleaning Works	Employment provider	Provides street cleaning service to supportive local authority: 75–100 per cent commercial income.	Free premises and grants from government to pay salary of chief executive	Providing homeless people with employment as a stepping stone to the mainstream labour market.
Environmental Training	Accredited training of homeless people	Wide range of projects, including accredited training program for homeless people: 75–100 per cent commercial income.	Volunteer teachers. Other volunteers help with environmental work.	Environmental goals—reducing carbon footprints. Work integration of homeless people a secondary social goal.
Garden Furniture Training	Training and work experience	Sale of handmade garden furniture: 0–25 per cent commercial income.	Free premises. Salaries paid by parent housing association. Grants to purchase materials	To provide homeless people with work experience in a supportive environment.
Integrated Living	Participation-based community	Sale of recycled goods. Accommodation provision: 50 per cent income derived through housing benefit; 25 per cent commercial income.	Free premises. Salaries of key staff paid by grants. Relies heavily upon volunteers and donations of goods.	To help homeless people reintegrate into society through 'work' and sheltered accommodation.

and selling garden furniture. Integrated Living offered a supportive working and living environment, which was portrayed as a re-integrative bridge into mainstream society. In each of these cases, interviewees presented the supportive working or training environment as more important than any specific qualifications or skills learned.

Three of the case study organizations' primary social goal was to provide paid employment to homeless people. Both Street Cleaning Works and Dry Cleaning Works offered a form of sheltered employment solely to homeless people. Employees were paid a regular monthly salary. This was intended as a transitional step to longer-term reintegration into the mainstream labour market. However, for some employees this was unlikely, and both organizations adopted a flexible approach dependent on individual circumstances. At Removal Works, following an initial training period homeless people were offered the chance of a permanent position. A paid position at Removal Works was thus presented as reintegration into the mainstream labour market.

Commercial considerations

All the case study organizations competed with private firms operating in their industrial fields. Environmental Training was paid according to the number of participants achieving qualifications, with an additional amount payable if participants moved into full-time employment, and again if they remained in a job for six months. All income was commercially generated. Integrated Living aimed to reduce homeless people's dependence on benefits, but was unable to generate sufficient commercial revenue from the sale of second-hand goods donated by members of the public, and so drew upon housing benefit to supplement trading income. Garden Furniture Training derived income through grants to pay (non-homeless) staff, and the sale of garden furniture that was used to cover other expenses.

The case study organizations providing paid employment to homeless people appeared more commercially orientated. Each derived the majority of their revenue through sales of goods or services in the marketplace. Street Cleaning Works derived most of their income through a contract with the local authority to provide street cleaning services. The salary of the chief executive and part of the rent for premises was covered by a grant provided by central government. Removal Works derived the majority of income through contracts and one-off payments from commercial companies to provide office removals. Salaries of those managing the enterprise were paid by the parent charity. Dry cleaning services derived most of their income through payments from employees of commercial companies for dry cleaning services. Items were collected and returned by homeless people from the commercial premises.

Each of the case study organizations faced tensions between social and commercial considerations. The next two sections outline the different paths taken by the case studies to negotiate these tensions, which are summarized in Table 2. Firstly, attention is paid to balancing strategies used by the case study organizations to negotiate inherent tensions between social and commercial considerations. To some extent, each organization adopted a more commercial focus in line with the industrial field in which they operated. Secondly, however, a number of conscious and unconscious resource transfer strategies were used that drew in resources from other sectors of the economy and mitigated any reduction in social outputs. This enabled some of the case study organizations to move beyond the limits of aggregate social and commercial outputs implied by the model.

Table 2. Approaches to negotiating social and commercial goals

Case study	Commercial income (per cent)	Relative priority paid to social or commercial goals	Resource transfer strategies
Dry Cleaning Works	50–75	Pro-social: Resisted commercial pressures to employ people on flexible contracts.	Moderate hybridity: Resource transfers from consumer, private firms, and foundations. In-house social support.
Removal Works	75–100	Commercial: $\frac{3}{4}$ of employees had never been homeless. Employees on flexible contracts. Participants must be free from substance misuse problems and be living in settled accommodation.	Low hybridity: Resource transfers from consumer, private firms, and relied on third sector to provide social support.
Street Cleaning Works	75–100	Pro-social: Employees paid above minimum wage on full-time contracts. Aimed to reduce time participants must be free from substance misuse.	Moderate hybridity: Resource transfers from consumer, private firms, government grants. In-house social support.
Environmental Training	75–100	Balanced: Admitted creaming off, but counter-balanced this by offering the same training activities to other homeless people using profits from other programmes.	Low hybridity: Resource transfers from consumer, volunteers. Relied on third sector to provide social support.
Garden Furniture Training	0–25	Pro-social: Limited creaming off.	High hybridity: Resource transfers from consumer, third sector, and state benefits system. Relied on third sector to provide social support.
Integrated Living	0–25 (+50 per cent housing benefit)	Pro-social: Limited creaming off. Excluded participants with substance misuse or moderate to severe mental health problems.	High hybridity: Resource transfers from consumer, third sector, state benefits system, volunteers and philanthropy. Mix of in-house and externally provided social support.

Findings

Balancing Strategies

Each case study organization offering paid employment to homeless people recognized the additional costs of employing disadvantaged groups, in part not only due to the cost of providing social support, but also due to homeless people being less productive than labour sourced from the open market. For example, when Removal Works was first established the intention was for all employees to be homeless. However, this had caused some problems:

> The problems were around what you'd imagine … inappropriate language, we work in a corporate setting, it's not a building site, there was a lot of street language, spitting, smoking where they shouldn't have been … one of them offering a security guard out for a fight, and you're just, like, oh my God! … just completely not understanding the environment they're working … every job we went out, something would happen, and it was only the good grace of the clients that we were working with that kept us going … and they would just use us sparingly, but at least they were throwing work at us, because they believed in what we were trying to achieve … that's what happened in the early days and that's why we stopped using 100 per cent ex homeless [people]. (Removal Works)

So, balancing the social and commercial goals for Removal Works involved compromising the original social mission to increase productivity.

Adopting the commercial practices of the field. In line with the commercial removals industry in which Removal Works operated, employees state which hours they are available to work, and if work is available are called in. If Removal Works won a contract, they were able to draw upon a pool of labour to fulfil the contract while keeping employment costs low. This led to considerable variation in hours worked and associated income for homeless people.

Dry Cleaning Works provided a service to individuals working at corporations in the commercial district. They had received much critical acclaim as a social enterprise, and had been visited by members of the government and Royal family. Dry Cleaning Works provided all employees with full- or part-time employment contracts. If they did not bring in commercial revenue, it was the organization that lost money rather than homeless employees. While this was a more pro-social approach, when the recession arrived, they were unable to reduce costs because of higher fixed overheads and went into receivership. A follow-up interview with the former chief executive revealed he had considered switching to offering homeless people employment on a self-employed basis. However, he felt that this went against the ethos of Dry Cleaning Works and risked exploiting homeless people.

These cases would suggest that some social enterprises have to adapt to employment conditions in the field within which they operate to survive. This may run counter to the social goals of the organization. However, other things being equal, not doing so could threaten organizational survival.

Creaming off. The organizations in this study all faced inherent pressure to cream off those homeless people easiest to place into the labour market. Each social enterprise engaged in creaming off, although the extent varied by organization. For example, Integrated Living recognized that their community was not suitable for all homeless people, particularly those with moderate to severe mental health problems and/or substance misuse problems, and actively discouraged these people from taking part. The other organizations also insisted that participants should be free from substance misuse. This is perhaps necessary for businesses facing commercial pressures to sell goods and services, not only because of potential adverse impressions presented to consumers, but also because of the effect on other participants.

However, the case study organizations also recognized the complexity of these competing commercial and social logics. Two, in particular, sought to redress the balance. Street Cleaning Works had been gradually reducing the period of time they expected employees to remain free from substance misuse problems. The chief executive recognized the dangers in this, but felt that if they were to pursue their social agenda they needed to reduce this period, and if necessary draw upon other income streams to facilitate this.

Environmental Training also recognized the incentive to cream off clients easiest to place into the labour market. They were paid for each homeless person moved into employment, with no distinction made between the varying levels of social need experienced by different homeless people (Stephens & Fitzpatrick, 2007). From a commercial perspective, it made sense to take on only those homeless people who were 'employment ready'. Their 'solution' was to accept participants who passed an initial screening test. However, people who 'failed' this screening test were offered the opportunity to take part informally in the various activities offered. This was effectively cross-subsidized by more profitable activities in other parts of the organization.

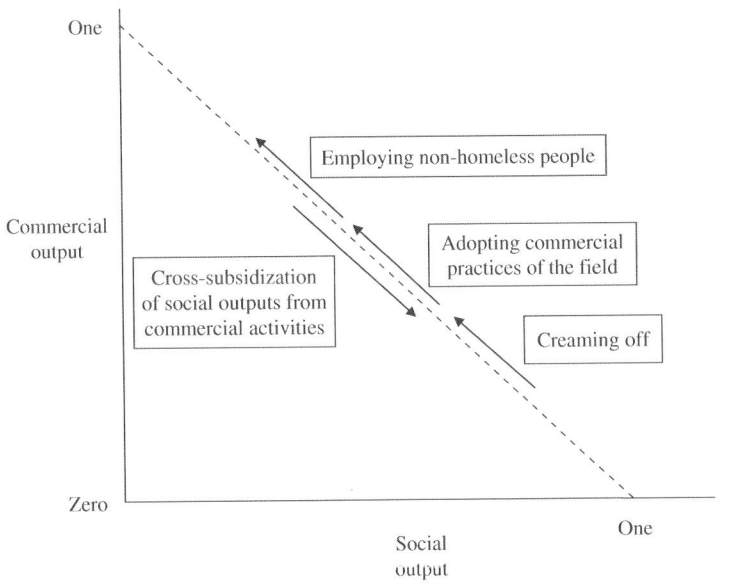

Figure 2. Strategies used to balance social and commercial considerations.

Social enterprises, particularly training and employment providers paid on an outcomes basis, face institutional pressures that push them towards the commercial pole of the model (see Figure 2). Indeed, other things being equal, to survive as enterprises they will need to mimic the behaviour of private firms in their industries. The case of Environmental Training shows that more socially orientated organizations may seek to subsidize social outputs from more commercially successful elements of their business. Although the other social enterprises in this study did not have other commercial arms from which to subsidize their more social outputs, they were able to transfer costs to other sectors of the economy.

Resource Transfer Strategies

Drawing upon hybrid resource mixes. Given that only Environmental Training had been able to rely solely on trading income in the previous year, it might be perceived that the case study organizations were not successful social enterprises, at least as laid out in much of the academic and policy literature. Interviewees were aware of this discourse of financial sustainability, and felt they were expected to aspire to this goal.

> First and foremost, we're a business and we have to be competitive in the marketplace, we have to deliver an excellent service, we have to be priced competitively, we have to be able to deliver in competition with all the other companies out there . . . . (Removal Works)

However, following probing as to the desirability of this end-goal, respondents all suggested that they were comfortable with a more hybrid resource mix whereby income was derived through a range of sources, including gifts in kind, grants and donations, and volunteer labour. For example, the manager of an Integrated Living project emphasized that the 'business' was financially sustainable; she also recognized that it could not survive without donations of goods and money from the local community, housing benefit payments to pay for the accommodation of homeless people living there, grants to refurbish the buildings, and the voluntary labour committed by local people to help manage the shops and recycling facilities. Trading income was just a minority element of the resource mix used to sustain Integrated Living.

While other social enterprises in this study drew upon a different mix of resources, each relied to some extent on grants, donations in kind and/or voluntary labour. This can be seen as reflecting their legitimacy as third sector organizations, and relative embeddedness within that sector (Di Maggio & Anheier, 1990). Other things being equal, this ability to draw upon resources available to third sector organizations provided social enterprises with a competitive advantage over private sector competitors.

Resource transfer from the consumer. While third sector organizations draw upon their legitimacy to attract philanthropic donations and voluntary labour, private firms use their legitimacy to attract people to purchase their products (Suchman, 1995). Social enterprises may be able to use their legitimacy to blend these sources of income. To some extent, the social enterprises in this study were able to increase demand for their products at a given price by using the fact that their products were created or sold by homeless people as a unique selling point. Each relied on ethical consumers purchasing their products (see

Hibbert *et al.*, 2005). In the case of Garden Furniture Training this element of their business model was essential to their future survival.

> I know that one of our BENCHES will never compare to one that's been imported [from] India, but if we can get within 20 quid of it, 25 quid of it, then we can turn round and it can be almost like a—'you can choose this BENCH, and it's been made by homeless people'—and that's almost like the kind of fair trade mark... because I know for a fact that if I was buying a new BENCH and there was one that was 25 quid dearer but it was made by homeless people, I know that I would quite happily pay that. (Garden Furniture Training)

Ethical consumers need not only be private individuals. By virtue of a contract with a supportive local authority that recognized the wider social value of employing homeless people, Street Cleaning Works was able to provide homeless people with employment. The additional costs associated with employing homeless people were effectively passed on to the consumer—in this case the local authority. Street Cleaning Works knew that other things being equal local authorities would prefer them as a provider as they were able to demonstrate their social outputs of employing homeless people in a city with a visible homelessness problem. The local authority was prepared to pay a slightly higher price than might have been achieved if procurement had been on the basis of 'best price'. Thus, ethical consumption can be seen as an implicit subsidy from the consumer to help offset the additional costs of employing homeless people. Other things being equal, by relying on ethical consumption social enterprises are able to achieve a competitive advantage over private sector competitors.

Resource transfer from private firms. Much of the emphasis to date has been on factors that enable social enterprises to compete with private sector providers. The temptation is to present the private sector as the corporate beast against which social enterprises must compete or die. This neglects an important aspect, namely the degree of support provided by private sector businesses to the social enterprises in this study. Dry Cleaning Works before its demise relied heavily upon the goodwill of private sector companies, allowing them to offer services from their premises free of charge. Similarly Street Cleaning Works had been offered free premises from which to relocate by a local company. In both cases, the private businesses concerned were able to boost their CSR profile through being associated with social enterprises.

Removal Works and Dry Cleaning Works were both provided with start-up assistance by mainstream entrepreneurs keen to utilize their business skills for a social purpose. In neither case, did the entrepreneur seek a financial return on their investment. In the case of Removal Works, the entrepreneur who operated in the same industry, albeit on a bigger scale, lent them equipment to help start up, and continued to subcontract some smaller jobs to them. So, as well as being in competition with private sector providers, some social enterprises were able to draw upon private sector providers (even in the same field) for advice and the provision of resources.

Resource transfer from the state. Those social enterprises deriving a higher proportion of commercial revenue (Street Cleaning Works and Removal Works) were able to provide paid employment to homeless people. Those relying more on other income sources

(Garden Furniture Training and Integrated Living) provided training and work experience. It is plausible that more socially orientated organizations deriving income from a wider range of sources believe that training homeless people is of greater social value than employing them. It is also plausible that the homeless people involved with Garden Furniture Training and Integrated Living had more complex social problems and were further from the labour market than those employed by the other social enterprises.

Nonetheless, I would argue it is unlikely that the businesses run by Garden Furniture Training and Integrated Living could achieve sufficient commercial revenue to pay people a living wage. Instead, both organizations relied on housing benefit payments, either explicitly or implicitly. Integrated Living's business plan relied upon the payment of housing benefit to people 'working' there. Employees were paid a nominal 'wage' of £30 a week—which did not affect their housing benefit, and was structured outside of the national insurance system. If Integrated Living had paid homeless people at the minimum wage, the housing benefit they received would have been reduced, so withdrawing a key financial element of their business plan. At the same time, the social enterprise would have increased the costs necessary to sustain themselves. Similarly Garden Furniture Training's trainees were predominately housed in hostels run by their parent housing association. Paying homeless people a wage would have adverse financial implications for their parent.

Relying on the tax and benefits system to increase the economic value of a job to employees is of course common in the private sector. According to GLA Economics (2011), the minimum wage necessary for a full-time single worker in London to meet their basic living costs in 2011 was £7.55 per hour. Around 10 per cent of full-time single workers receive less than this amount, but many avoid absolute poverty through receipt of means tested benefits. However, as one interviewee from this study recognized, most private sector businesses are unable to draw upon free labour:

> I think some of these charities who have been encouraged to explore social business through things like Places for Change ... they don't actually employ their staff, they use a voluntary training programme ... although they are generating money ... they're not actually creating employment. Just opportunity and experience. Which is valuable, but it's not a business, because businesses don't have free workers [laughing]. (Dry Cleaning Works)

This ability to draw upon free labour effectively involves a transfer of resources from the state to the social enterprise. Without the housing benefit system, the training and work experience providers in this study would not have been financially viable. By drawing upon these state resources, other things being equal social enterprises are able to gain a competitive advantage over private sector competitors in the same field.

Resource transfer from the third sector. Embedded work integration social enterprises in the homelessness field may find it difficult to derive sufficient commercial revenue to provide social support to employees. There may be a tendency for voluntary organizations moving towards a social enterprise model to stop providing social support to their clients if this provision is not cost effective (Dart, 2004; Russell & Scott, 2007). None of the organizations in this study were able to fund social support through commercial revenue. Instead, three broad approaches to providing social support were identified.

Removal Works described themselves first and foremost as a business. They argued it was not their job to provide social support to clients. Instead, they felt that this was best left to other organizations with expertise in this area. This left them free to focus on being a business that employed formerly homeless people. This approach was also followed by Garden Furniture Training. In both cases, some social support to employees was provided by their parent charity/housing association.

Environmental Training and Integrated Living both offered informal social support to trainees. Where more professional advice and support was deemed necessary, both organizations referred homeless people to other (third sector) organizations with more expertise.

Both Dry Cleaning Works and Street Cleaning Works employed staff to provide social support and advice to homeless employees. In both cases, the costs of this were met by grants from foundations or government.

Therefore, none of the case study organizations was able to provide social support to clients funded through commercial revenue. Each relied on other organizations to provide or fund social support. This enabled them to avoid the financial collapse suffered by high-profile organizations in the homelessness field, such as Aspire, which had underestimated the additional support needs required by their homeless employees (Mozier & Tracey, 2010; Tracey & Jarvis, 2007). Rather than attempting to derive sufficient commercial surplus to provide social support, the case study organizations relied on other third sector organizations to fund or provide this support to their employees. This is effectively a resource transfer from the third sector to social enterprises.

Conclusions

This research adopted an exploratory qualitative case study approach to identify ways in which social enterprises in the homelessness field negotiate social and commercial considerations. To achieve some conceptual clarity, the main financial consideration studied was deriving sufficient trading revenue to remain in business. Social considerations were largely reduced to moving homeless people into employment. Of course, the wider picture is likely to be far more complex. Additionally, by interviewing key figures within each organization, the voices of homeless people themselves were omitted. Nonetheless, this study makes an important contribution to theory building.

Work integration social enterprises in the homelessness field face inherent tensions in their business model between social and economic objectives. Many homeless people that they seek to help into employment are less economically productive than employees hired by competitors in the private sector, and may require additional social support. The nature of their funding environment means they face institutional incentives to cream off those easiest to place into employment, or to adopt the employment practices of the field in which they operate. However, this may run contrary to their social aims and values. It would seem that organizations deriving a higher proportion of commercial revenue are more likely to be pulled towards the commercial pole of the model, at the expense of social goals. Other things being equal, the social enterprises in this study faced an inherent trade-off between social and commercial considerations.

Some social enterprises in this study were able to adopt strategies to resist these institutional pressures. These involved drawing upon their hybrid nature to access resources from other sectors of the economy. These resource transfers allow individual

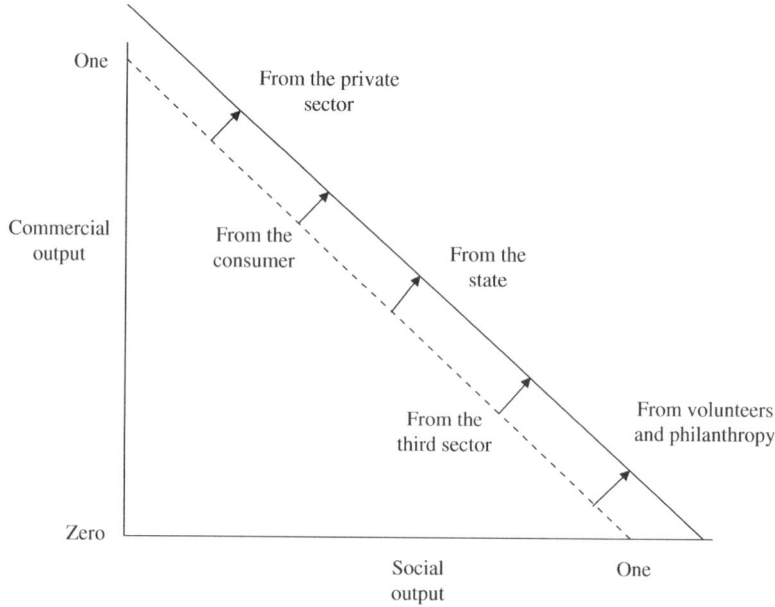

Figure 3. Resource transfer strategies used to subsidise social and commercial outputs.

social enterprises to move beyond the constraints of aggregate social and commercial output implied by the model developed in this paper (see Figure 3). Thus, social enterprises were able to draw upon resources available to charities—private donations and voluntary labour. Additionally, the social enterprises in this study were able to utilize their ethical dimensions to draw in resources from consumers. Further (direct or indirect) resource transfers were often obtained from other private businesses in similar fields. By providing 'training' rather than direct employment, some social enterprises receive resource transfers from government; firstly through payments for people moved into employment, and secondly through relying on the state to meet the living costs of homeless people. Social enterprises may rely upon other charities to provide social support to homeless people, or draw in funding from elsewhere to provide this support in-house.

These strategies involved the subsidization of social enterprise by volunteers, funders, consumers, the state benefit system, and other charities or homeless employees working for less than the minimum wage. None of the organizations in this study were able to rely wholly on commercial revenue to move people into employment and provide wider social support. As outlined in the earlier sections of this paper, there is a policy assumption in England that social enterprises successfully combine social and commercial goals (DTI, 2002; OTS, 2006). The assumption that social and commercial goals can be successfully combined within an embedded social enterprise implies that prioritizing profit maximization maximizes social benefits. However, this study suggests a policy assumption that social enterprises can address homelessness while generating a surplus from their trading activities should be treated with some caution. Instead, the social enterprises in this study, which at first sight would appear embedded social enterprises, should be conceptualized as hybrid organizations able to draw upon a wider range of resources than mainstream businesses or traditional charities. This is because they are able to blend the legitimacy of third sector organizations with the legitimacy of private firms through balancing the institutional logics of both. While these additional resources might be 'free' to the social enterprise, they involve redistribution from the state, individual, private firms, and existing charities. Successfully attracting and maintaining these different resources is likely to require careful management. The extent to which balancing a hybrid resource mix while simultaneously negotiating competing social and commercial logics is sustainable for any single organization in the longer term is questionable, particularly in an environment that valorizes market forces over and above social goals.

Acknowledgements

The author would like to thank three anonymous referees and the editor of the journal for valuable suggestions to improve the paper. Additionally, the author would like to thank Gemma McKenna, David Mullins, and Nina Teasdale for comments on early drafts. This paper derives from a programme of work being carried out at Birmingham and Southampton Universities as part of the Third Sector Research Centre (TSRC). The support of the Economic and Social Research Council (ESRC), the Office of Civil Society (OCS), and the Barrow Cadbury UK Trust is gratefully acknowledged.

References

Aiken, M. (2006) Towards market or state: tensions and opportunities in the evolutionary path of three types of UK Social Enterprise, in: M. Nyssens (Ed.) *Social Enterprise: At the Crossroads of Market, Public Policies and Civil Society*, pp. 259–271 (London: Routledge).

Aiken, M. & Bode, I. (2009) Killing the golden goose? Third sector organizations and back to work programmes in Germany and the UK, *Social Policy and Administration*, 43(3), pp. 209–225.

Alter, K. (2007) *Social Enterprise Typology* (Portland: Virtue Ventures LLC).

Amin, A., Cameron, A. & Hudson, R. (2002) *Placing the Social Economy* (Oxford: Routledge).

Billis, D. (2010) Towards a theory of hybrid organizations, in: D. Billis (Ed.) *Hybrid Organizations and the Third Sector*, pp. 46–69 (London: Palgrave).

Blackburn, R. & Ram, M. (2006) Fix or fixation? The contributions and limitations of entrepreneurship and small firms to combating social exclusion, *Entrepreneurship and Regional Development*, 18(1), pp. 73–89.

Bornstein, D. (2001) *How to Change the World: Social Entrepreneurs and the Power of New Ideas* (Oxford: Oxford University Press).

Carmel, E. & Harlock, J. (2008) Instituting the 'third sector' as a governable terrain: partnership, performance and procurement in the UK, *Policy and Politics*, 36(2), pp. 155–171.

CLG (Communities and Local Government) (2006) *Places of Change: Tackling Homelessness Through the Hostels Capital Improvement Programme* (London: CLG).

CLG (2007) *Creating Places of Change: Lessons Learnt from the Hostels Capital Improvement Programme 2005–2008* (London: CLG).

Cornforth, C., Thomas, A., Spear, R. & Lewis, J. (1988) *Developing Successful Worker Co-Operatives* (London: Sage).

Dart, R. (2004) Being 'business-like' in a nonprofit organization: a grounded and inductive typology, *Nonprofit and Voluntary Sector Quarterly*, 33(2), pp. 290–310.

Dees, J. G. (1998) Enterprising nonprofits, *Harvard Business Review*, January/February 1998, pp. 55–67.

Dees, J. G. (2007) Taking social entrepreneurship seriously, *Society*, 44(3), pp. 24–31.

Defourny, J. & Nyssens, M. (2010) Conceptions of social enterprise and social entrepreneurship in Europe and the United States: convergences and differences, *Journal of Social Entrepreneurship*, 1(1), pp. 32–53.

Dey, P. & Steyaert, C. (2010) The politics of narrating social entrepreneurship, *Journal of Enterprising Communities*, 1(4), pp. 85–108.

Di Maggio, P. J. & Anheier, H. K. (1990) The sociology of nonprofit organizations and sectors, *Annual Review of Sociology*, 16, pp. 137–159.

DTI (Department of Trade and Industry) (2002) *Social Enterprise: A Strategy for Success* (London: DTI).

Friedman, M. (1970) The social responsibility of business is to increase its profits, *New York Times Magazine*, September 13.

Gardin, L. (2006) A variety of resource mixes within social enterprises, in: M. Nyssens (Ed.) *Social Enterprise: At the Crossroads of Market, Public Policies and Civil Society*, pp. 111–136 (London: Routledge).

GLA (Greater London Authority) Economics (2011) *A fairer London: The 2011 Living Wage in London* (London: GLA).

Hansman, H. (1987) Economic Theories of Nonprofit Organization, in: W. Powell (ed), *The Nonprofit Sector: A Research Handbook*, pp. 27–42 (New Haven, CT: Yale University Press).

Hibbert, S., Hogg, G. & Quinn, T. (2005) Social entrepreneurship: understanding consumer motives for buying The Big Issue, *Journal of Consumer Behaviour*, 4(3), pp. 159–172.

Kerlin, J. (2010) A comparative analysis of the global emergence of social enterprise, *Voluntas*, 21(2), pp. 162–179.

Lyon, F., Teasdale, S. & Baldock, R. (2010) *Approaches to Measuring the Scale of the Social Enterprise Sector in the UK*. TSRC Working Paper 43 (Birmingham: University of Birmingham).

McKenna, G. (2011) The UK's next top model: the ambiguities and complexities of social enterprise models in the homelessness sector. Paper presented at the Housing Studies Association Conference, University of York, April 14.

Mozier, J. & Tracey, P. (2010) Strategy making in social enterprise: the role of resource allocation and its effects on organizational sustainability, *Systems Research and Behavioral Science*, 27, pp. 252–266.

Mullins, D. & Riseborough, M. (2000) *What are Housing Associations Becoming? Final Report of Changing with Times Project*. Housing Research at CURS 7 (Birmingham: University of Birmingham).

Nicholls, A. (2010) Institutionalizing social entrepreneurship in regulatory space: reporting and disclosure by Community Interest Companies, *Accounting, Organizations and Society*, 35(4), pp. 394–415.

Nyssens, M. (Ed.) (2006) *Social Enterprise: At the Crossroads of Market, Public Policies and Civil Society* (London: Routledge).

ODPM (Office of the Deputy Prime Minister) (2003) *Homelessness Statistics: June 2003* (London: ODPM).

ODPM (2005) *Hostels Capital Improvement Programme Policy Briefing 12* (London: ODPM).

OTS (Office of the Third Sector) (2006) *Social Enterprise Action Plan: Scaling New Heights* (London: OTS).

Peattie, K. & Morley, A. (2008) *Social Enterprises: Diversity, Dynamics, Contexts and Contributions* (Cardiff: ESRC/Social Enterprise Coalition).

Pleace, N. (1999) Single homelessness as social exclusion: The unique and the extreme, *Social Policy and Administration*, 32(1), pp. 46–59.

Russell, L. & Scott, D. (2007) *Social Enterprise in Practice* (Manchester: Charities Aid Foundation).

Short, J. C., Moss, T. W. & Lumpkin, G. T. (2009) Research in social entrepreneurship, past contributions and future opportunities, *Strategic Entrepreneurship Journal*, 3, pp. 161–194.

Spear, R. & Bidet, E. (2005) Social enterprise for work integration in 12 European countries: a descriptive analysis, *Annals of Public and Co-Operative Economics*, 67(2), pp. 195–231.

Stephens, M. & Fitzpatrick, S. (2007) Welfare regimes, housing systems and homelessness: how are they linked? *European Journal of Homelessness*, 1, pp. 201–211.

Suchman, M. C. (1995) Managing legitimacy: strategic and institutional approaches, *Academy of Management Review*, 20(3), pp. 571–610.

Teasdale, S. (2010a) Models of social enterprise in the homelessness field, *Social Enterprise Journal*, 6(1), pp. 23–34.

Teasdale, S. (2010b) How can social enterprise address disadvantage? Evidence from an inner city community, *Journal of Nonprofit & Public Sector Marketing*, 22(2), pp. 89–107.

Teasdale, S. (2011) What's in a name? Making sense of social enterprise discourses, *Public Policy and Administration*, published online 25 May 2011. doi: 10.1177/0952076711401466.

Tracey, P. & Jarvis, O. (2007) Toward a theory of social venture franchising, *Entrepreneurship, Theory and Practice*, 31(5), pp. 667–685.

Tracey, P., Phillips, N. & Jarvis, D. (2011) Bridging institutional entrepreneurship and the creation of new organizational forms: A multilevel model, *Organization Science*, 22(1), pp. 60–80.

Wright, I. (2008) Win a Share of £1.6 Million: Spark Challenge Now Open for Social Enterprises Tackling Homelessness. Available at http://www.sparkchallenge.org/news_10_12_08.html (accessed 19 June 2011).

Yin, R. (2003) *Case Study Research: Design and Methods (Third Edition)* (London: Sage).

Young, D. (2010) The state of theory and research on social enterprises. Paper presented at the Exploring Social Enterprise Conference, University College of Los Angeles, October 29.

Hybridity Enacted in a Large English Housing Association: A Tale of Strategy, Culture and Community Investment

HALIMA SACRANIE

Centre for Urban and Regional Studies, Birmingham Business School, University of Birmingham, Birmingham, UK

ABSTRACT   *This paper seeks to advance the understanding of hybridity in the social housing sector by drawing on a multi-layered case study of a single, large housing association X (HAX) to illustrate how the competing logics that underpin that hybridity are enacted at a small, locally based subsidiary (Small Housing Association). The inherent paradoxes and complexity that characterise the third sector of housing in the UK have been explored in a study of the changing strategic management and organisational culture for community investment activities over a 2-year period at HAX. The study links the concepts of institutional logics [Friedland & Alford (1991) The New Institutionalism in Organizational Analysis. (Chicago, IL: University of Chicago Press); Thornton & Ocasio (1999)* American Journal of Sociology, *105(3), pp. 801–843] in a social housing context [Mullins (2006)* Public Policy and Administration, *21(3), pp.6–21] with organisational cultures [Gregory (1983)* Administrative Science Quarterly, *28, pp. 359–376; Hofstede, 1993] to locate the strategic focus of the organisation in a logics–culture matrix. A link between a consumerist or customer logic and a prevailing corporate culture is identified, together with a more historic connection between a community logic and a weakening regional, locally responsive culture.*

Introduction

English housing associations provide a well-established example of hybrid organisations, operating with a mixture of logics derived from their roots and links to communities, the state and the market (Billis, 2010; Mullins & Pawson, 2010). As the largest providers of affordable rented housing, they provide around 2.5 million affordable rented homes for 5 million people from financially disadvantaged or socially vulnerable backgrounds (Housing Corporation, 2005). With banks tightening purse-strings and the government under pressure to reduce the national deficit through public sector reform, and with new challenges to community linkages from the Big Society and localism agendas (DCLG, 2010) in England, how these hybrid social housing organisations adapt and modify to meet divergent demands is both intriguing and topical.

This paper draws on a qualitative longitudinal case study, delving into the inner workings, organisational culture and multiple realities of a prominent housing association in the UK, named Housing Association X (HAX). During the study, HAX embarked on a large-scale restructuring programme, centralising its key functions and consolidating its staffing structure across its operating companies. One of the key developments was a new, centralised department for community investment (CI) which focused on developing an organisation-wide strategy for neighbourhood investment, financial inclusion and employment and training work formerly undertaken through more *ad hoc*, localised CI projects.

The broad aim of the longitudinal study was to explore the changing strategic management and organisational culture at HAX by tracking the development of its CI strategy over a 2-year period, as a reflection of multi-layered views and sub-cultures within the organisation (Gregory, 1983; Johnson, 1992; Schein, 1997) overlaid with driving institutional logics (Lounsbury, 2007; Mullins, 2006; Scott, 2001; Thornton, 2004).

Without wishing to preempt the discussion on theoretical concepts that follows in 'A Multi-layered Approach: Methodological and Theoretical Frameworks' section of this paper, a quick definition of the key concepts on organisational culture and institutional logics will help to the clarify the main themes in the backround to this research. According to Scott (2001, p. 139) 'Institutional logics refer to the belief systems and related practices that predominate in an organisational field'. A well-cited definition of organisational culture also makes reference to these intangible belief systems, with organisational culture regarded as the 'pattern of beliefs and expectations shared by the organisations members. These beliefs and expectations produce norms that powerfully shape the behaviour of individuals and groups in the organisation' (Shwartz & Davis, 1981, p. 33). Furthermore within a single organisation, multiple cultures can and do exist. Gregory (1983, p. 359) contends that:

> ... many organisations are most accurately viewed as multi-cultural. Sub-groups within different occupational, divisional, ethnic, or, other cultures approach organisational interactions with their own meanings, and senses of priorities ...

This study adopted a multi-layered approach on a number of dimensions: exploring the multiple layers or sub-cultures within the single case to reveal the different views, perceptions and co-existent realities within the organisation; and tying together methodology and theory by layering organisational sub-cultures with external and institutional factors. This multi-layered framework translated into the project's research design, by taking vertical and horizontal cross sections as sampling points through the management hierarchy and functional and geographic boundaries of the organisation.

Even from the author's earliest engagement with HAX in 2007, it became clear that the hybrid organisation itself represented a microcosm of the paradoxes that were evident at a sector level. These paradoxes were more pronounced because of the timing of the research, from 2007 to 2010, which coincided with a period of intense debate about the legal status, independence and regulation of housing associations in England (Cave, 2007; Mullins *et al.*, 2009; NHF, 2008b). It became crucial to the author to reflect these complexities in the underlying questions the case study would address, which were the following:

- How are housing associations changing?
- Why are housing associations changing in the way they are?
- How are the changing identity and cultures of these organisations manifest?
- What are the kind of change drivers that are bringing about these changes?

As a third sector social business with market, state and community priorities, a study exploring the changes in identity of a large housing association such as HAX, would essentially provide the opportunity to empirically capture hybridity in action.

The specific aim of this paper is to discuss this process of hybridity enacted by employing the theoretical perspectives of institutional logics and organisational culture to reflect on some findings from the broader case study. This paper focuses on the experiences of Small Housing Association (SHA), a subsidiary covered in depth as one layer of the overall project described above.

'Strategy, Culture and CI' section of this paper includes a brief sector overview highlighting some of the complexity and critical tensions manifest within the hybrid social housing sector in England, and in particular with reference to the key themes of 'Strategy', 'Culture' and 'CI'. The section also provides a background to the case study organisation HAX and the subsidiary association SHA. 'A Multi-layered Approach: Methodological and Theoretical Frameworks' section considers aspects of the methodology, research design and theoretical positioning of the wider study which informs the evidence presented in this paper. This empirical evidence on SHA is presented in 'Changes in CI at SHA' section, while 'Hybridity Enacted: Changing Organisational Culture and Institutional' Logics section develops these findings into an interpretative framework: the cultures/logics matrix. The last section concludes by drawing out the implications of the study design and findings for future work on the enactment of change between competing logics that characterise hybrid social organisations.

Strategy, Culture and CI

The English Housing Association Sector

The evolution of housing associations from a historic context to their current status as entrenched hybrid third sector organisations (Billis, 2010), performing a complicated balancing act between state, market and society, has been well documented (Gruis, 2009; Malpass, 2000; McDermont, 2010; Mullins & Riseborough, 2000; Mullins, 2006). These entrenched forms of hybridity have been defined by Billis (2010, p. 59) as the permanent influence by public and private actors on the governance and operations of an organisation in return for the resources provided by these actors. Major strategic shifts have occurred since the 1980s when Government policy and the liberalisation of private equity led to the adoption of mixed or hybrid funding models, leading to a greater commercial focus by these organisations in an expanding sector (Mullins & Riseborough, 2000). Over the last two decades, there has been a significant repositioning of the sector amidst retreating welfare states, regulatory pressures, accountability gaps, increasing business rationales and changing stakeholder relationships (Mullins et al., 2008).

Labour government policies such as Investment Partnering, which concentrated public funding on the largest housing associations, increasing costs of regulatory compliance (Cowan & McDermont, 2006; Mullins, 1997) and a quest to a achieve economies of scale, saw the rationalisation of the sector in a pre-recession increase in merger activity, which created mega organisations in the top echelons of the sector (Pawson & Sosenko, 2012). These large commercial organisations needed to strategically balance achieving efficiency targets and economies of scale with local service delivery and accountability (Mullins, 2006). However, the strategies adopted to accomplish this balancing act have not been homogenous, with a lack of consensus regarding the core function of housing associations, particularly with regard to diversification into private markets on the one hand, and performing social or community services formerly undertaken in the public realm on the other. Figure 1 summarises some of these divergent demands on the sector at the time this research was conducted.

Looking specifically at changing organisational cultures, as housing associations have evolved, there has been a shift in organisational culture from traditional organisational structures, and public and voluntary characteristics, towards rational, business cultures and forms of social entrepreneurship (Czischke *et al.*, 2010; Mullins & Riseborough, 2000). The pre-recession growth of mega housing associations with an expanding scale of operations and the increasing importance of private lenders as stakeholders have resulted in a growing corporate culture, with an emphasis on business models, corporate styles of strategic planning and increased financial reporting, similar to large private sector companies (Mulllins & Sacranie, 2008). Governing boards have become more streamlined and professional, with payment for non-executive directors since 2004, and the increasing dominance of executives on governing boards, reflecting the growing 'managerialisation' (Walker, 2000), 'governing independence and expertise' (McDermont, 2010) of these housing associations.

CI is a key focus for this paper and indeed the broader case study from which it draws on. It can be argued that the retreating welfare state has generated expectations for housing

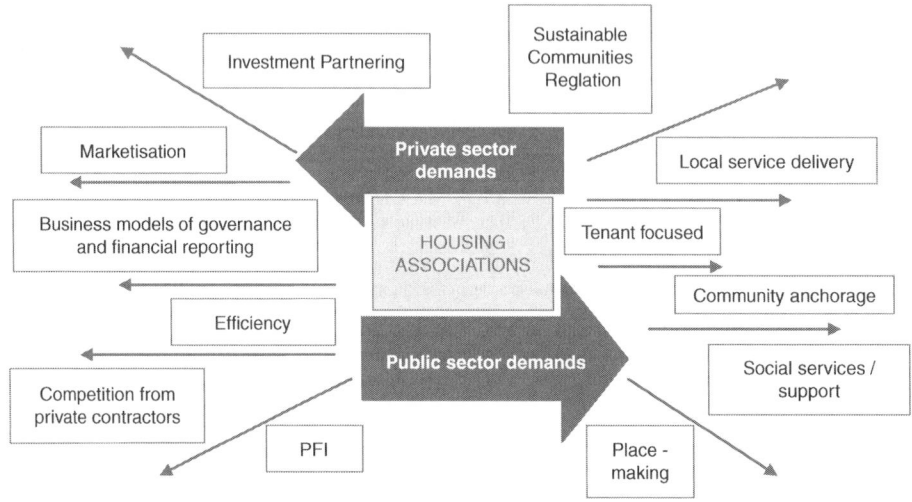

Figure 1. Divergent demands on social housing organisations. *Source*: The Author (2008b).

associations to fulfil wider community and social welfare obligations beyond their core housing role (Brandsen *et al.*, 2006; Heino *et al.*, 2007). In the English context, as the sector has grown employing mixed funding models, housing associations activities have expanded to include social welfare type or CI services to complement their core housing management role, variously known as 'housing plus', a 'wider role' and 'in Business for Neighbourhoods' (Clapham & Evans 1998; Mullins *et al.*, 2001; National Housing Federation 2003, 2008a, b).

According to the National Housing Federation's first national audit of neighbourhood services and facilities, housing associations provided 6800 neighbourhood services and either provided or managed hundreds of neighbourhood facilities, which benefited around 5.5 million people (NHF, 2008a). Over the period of the audit, the total CI spend was around £435 million, of which £272 was provided by the housing associations themselves and the balance externally sourced. The key areas of investment were neighbourhood facilities, safety and cohesion, environmental services, education and skills and employment and enterprise (NHF, 2008a).

This audit took place around the same time as intense debate in the sector around the formation of the (then) new regulatory body, the Tenant Services Authority. This debate questioned to what extent these CI activities would be externally regulated by the government (DCLG, 2007). Inherent to both the philanthropic and local authority, legacy of some housing associations was a sense of civic duty and accountability to tenants and communities. However, arguments presented in the Cave Review evidence (2007) revealed a divergence of views in relation to what constitutes 'housing plus' (Clapham & Evans, 1998) activities, or activities beyond the core function of providing and managing social rented accommodation. The burden of responsibility for neighbourhood investment and some community social services regarded as local authority responsibilities did emerge as a contentious issue. Resentment towards 'policy passporting' by the regulator, being set agendas by local and national government, taking on responsibility for non-core services and the possibility of being regulated for these resulted in the National Housing Federation's campaign in defence of third sector identity and independence prior to Housing and Regeneration Act being passed in July 2008.

With regard to legislative proposals for the regulation of non-core or CI activities, the National Housing's Federations strong response was:

> Housing associations do not do this (social work) because a housing regulator told them to. They do it because they see the local need and work with tenants and communities to meet that need, and

> Funders, such as charities and local partnerships, agree what they expect to see in return for their funding. They will not want the housing regulator to step in and steer such work. If it could do so, it could divert associations to concentrate on the 'flavour of the month' or what politicians and civil servants think will work. (NHF, 2008b)

As the above quotation suggests, the area of CI in social housing is one highly susceptible to debate as it reflects tensions between social, economic and political or public policy priorities. This tension is exacerbated by the blurred boundaries between what is private, public and third sector within the social housing world. For this reason, the author considered CI a particularly useful focus to explore the hybrid nature of a large housing association. Changing approaches to the strategic management and delivery of CI, as driven by organisational culture and instititutional logics, could serve as an example of hybridity in action in this case of HAX.

HAX: A Case Study of a Large Housing Association

As one of the largest housing associations in England, HAX is made up of three main operating companies, as well as a number of smaller, niche subsidiaries (such as SHA), and a maintenance and repairs service. HAX owns and manages a housing stock of over 50 000 homes, in over 80 Local Authorities across England including major towns and cities, as well as in some rural areas in the South West, East, Midlands and North Regions. Its development programme provides up to 1000 new rented and low-cost home ownership properties (since 2008) in over 30 of these Local Authorities areas throughout the country.

The bulk of housing stock at HAX is owned and managed through its three large operating companies, which have come together as a group under the umbrella parent organisation HAX through a series of mergers over the last 10 years. The history of HAX is therefore the collective history of its operating companies and subsidiaries, as well as subsequent changes and developments at HAX as a unified organisation, which is the area this study explored.

At HAX, a strong focus on the strategy-making process was evident with corporate objectives, strategic goals and 'golden threads' linking group centre strategies with corporate planning at all levels of management across the operating companies and subsidiaries. This strong focus on strategic management was of particular relevance in the post-merger context the organisation was operating in. Senior executives at the organisation also acknowledged the impact of external drivers such as the regulatory debate and recessionary pressures on the strategic direction of the housing association.

As an organisation, HAX has a strong corporate outlook, driven by a business-oriented group executive leadership, with clear cultural variations vertically and horizontally across the management hierarchy and geographical stretch of the organisation. Management research on organisational culture and sub-cultures conducted by HAX in 2008 revealed that there were critical tensions between the different sub-cultures in the post-merger HAX, which brought together different companies with different heritage and cultures from across England. A wide perception across senior management, endorsed by frontline staff, was that a cultural fit or cohesion had not been achieved between the operating companies, for example with regard to protecting the heritage and identity of the individual operating companies, or the way in which staff felt local services should be delivered. Over the period of this case study research from 2007 to 2010, executive management at HAX was hoping to develop more of a cultural homogeneity within the organisation around a common HAX brand.

In fieldwork conducted at HAX, it was evident that there was a significant variance across the organisation as to what exactly constituted CI and how it should be framed. At the Group Executive level, there was a strong sense of financial and corporate purpose with the core function of HAX seen to be providing and managing affordable housing. Community support services were more often regarded as discretionary, part of a broader corporate social responsibility (CSR) agenda and appropriate where they complemented some of the financial or strategic aims of the organisation. Further down the management hierarchy, such as regional managers and frontline housing staff, there was a greater emphasis of the organisation's role and responsibilities to the wider community. At the same time within the senior executive team, there was a growing acknowledgement of the imbalance between a dominant financial purpose and efficiency agenda at HAX, and shortcomings in providing the 'softer stuff' such as CI. The perceived problem was that locally dispersed, sporadic services were being provided but with inconsistencies, duplication and inefficiencies according to senior management. Nevertheless, CI was becoming more important as the financially secure HAX aspired to develop its 'softer, feel-good' areas, necessary to be a key player in the industry, and so in effect 'playing catch-up' with the rest of the sector.

Over the course of the study, the regional CI teams were rationalised into a new centralised CI department to give a strategic focus and consistency to the previously localised, *ad hoc* services. The changes in the strategic direction of CI at HAX, and its link to questions about culture, accountability and the core role of HAX as a social landlord provided an ideal opportunity to investigate the changing identity of this hybrid organisation.

SHA: A Specialist Subsidiary

A focus within the broader study was a small, stand-alone black and minority ethnic (BME) subsidiary housing association within the parent organisation HAX. This association, which is referred to as SHA, while untypical of the larger income-generating operating companies in HAX, was nevertheless a pertinent example of the tensions or rival logics manifest in the broader case study. In its changing emergent and realised CI strategies, SHA reflects a shift in organisational culture from a local, regional focus to a more centralised, corporate vision.

To understand the context of SHA, it is useful to briefly consider the background of BME organisations. BME organisations originated in the 1970s to address the particular needs of ethnic minority communities which had, as a result of poverty and discrimination, poor access to housing (Mullins, 2010). A defining characteristic of a BME association was a governing Board with a majority of BME members. Furthermore, these organisations sought to not only improve access to housing for the specific communities they served but also fulfil a greater community role in advancing training and employment opportunities. In the 1980s, around 60 BME associations were supported by the Housing Corporation, and by 2003 these associations managed around 25 000 homes (Harrison, 1995; Mullins, 2010). Since then market pressures that have stimulated merger and consolidation activity in the social housing sector, as well a changing funding and policy climate, have led to BME organisations joining mergers or becoming part of group structures (Pawson & Sosenko, 2012). SHA was no exception to this trend, joining one of the large housing associations in 2001 that eventually became part of the HAX group in a merger in 2007.

SHA originated as a small BME housing association that targeted the housing and social needs of the South Asian community in a mid-sized English city. Particular problems for the community were seen to be overcrowding, poor housing stock in need of major repairs, and barriers of language and culture to accessing mainstream local authority services and advocacy. It was felt that a specific housing-related CI service should be set up to serve that particular community and a bid for external Lottery funding was won to develop a long-term (3–6 years) project. In the latter stages of the scheme, some internal HAX funding was set aside to supplement the external grant.

The purpose of SHA's CI programme was to provide a comprehensive housing advice service to the South Asian community, enabling the most disadvantaged within that ethnic minority group to access information, guidance and support in all matters relating to housing. This included providing advice to this community about the role of housing associations, which are not traditionally considered by South Asians in the UK as one of the tenure options available to meet their housing needs. The service was delivered through outreach advice surgeries, home visits, a confidential helpline and case workers attending appointments with clients. Originally planned to be a sign-posting agency, it developed into a more comprehensive service. At the time of the research, the programme provided in-depth casework according to the need of service users who required advocacy, assistance at appointments, filing cases for appeals, transfers, reviews, assisting with homelessness and emergency accommodation. The outreach service was popular with the South Asian community with over 3000 enquiries and in-depth cases dealt with over the 6-year duration of the project, which was awarded a number of external commendations.

As described earlier, the key changes in the strategic direction of CI at HAX had emerged in the form of a dramatic organisational restructure, with a number of regional departments being centralised functionally, one of which most noticeably being the formation of a new centralised CI department and team. In terms of direct impacts on SHA as a result of the general restructuring programme at HAX, being a rather niche and small association SHA, experienced fewer staffing changes than the larger operating companies at HAX. However, the changes in its CI programme reflect notable shifts in sub-cultures and dominant institutional logics which were typified by the organisation HAX as a whole. These changes and their implications for the changing identity of both SHA and HAX will be considered in detail in 'Changes in CI at SHA' and 'Hybridity Enacted: Changing Organisational Culture and Institutional Logics' sections of this paper.

A Multi-layered Approach: Methodological and Theoretical Frameworks

Using key themes identified by the author in a sector analysis conducted from 2007 to 2008, and preliminary fieldwork undertaken at HAX over the same period, the focus of the research enquiry on 'Strategy, Culture and CI' was fixed. At this stage, certain assumptions were incorporated as part of the theoretical framework for this project, namely conceptualising strategy as an emergent process (Mintzberg & Waters, 1985) and the concept of organisational cultural pardigms, sub-cultures or multiple identities revealing shared groups of characteristics within organisations (Gregory, 1983; Hofstede, 1998; Schein, 1997).

Seminal, well-cited theories were deliberately selected to provide the theoretical assumptions by which to develop research questions and develop topic guides around the key emerging issues both at a macro and organisational level. For example, exploring sub-cultures through a multi-layered study (Gregory, 1983; Hofstede, 1998) and identifying key components of organsiational culture such as organisational structure which became very relevant in the central versus regional tensions uncovered at HAX. These theoretical assumptions would provide the tools and terminology by which to undertake the inductive research, with the intention to eventually overlay these theories with more contemporary literature by means of developing theory further.

Within the broader case study framework, five layers of research were conducted as a multiple sampling-point strategy across a cross section of the management hierarchy at HAX, from board and senior executive level, through to frontline services and examples of

CI projects. The total number of interviews conducted were 62 and there were four focus group discussions, in cities and towns in the Midlands, North East, South East and South West of England. The majority of participants and interviewees were staff or employees at the case study organisation, ranging from the Chief Executive and Directors to senior, middle and frontline managers and housing officers. Focus groups included tenants and community members.

SHA, the focus for this paper, was the fourth of the case study layers and the fieldwork conducted at this subsidiary housing association included repeated interviews, a focus group discussion, visits and observation of the CI activities it undertook with South Asian communities in a mid-sized English city. Table 1 provides a more detailed outline of research methods for this part of the study.

A grounded approach was adopted for the data analysis (Glaser & Strauss, 1967). Fieldwork undertaken at HAX revealed that there was a clearly pronounced tension between a more corporate, financially driven and centralist outlook which dominated senior executive decision making and a more locally responsive, community-oriented, regional culture that was being displaced by the former. These seemed to broadly link with Mullins' (2006) competing logics of efficiency and accountability, which itself drew on the earlier work on multiple institutional logics (Scott, 2001; Thornton, 2004). Subsequently, this research sought to identify dominant sub-cultures and link these with key institutional logics within the multiple layers of the study. In this way, institutional logics were incorporated into the theoretical framework of the research to better understand the strategic direction of CI at HAX, and broadly to build on existing literature regarding the future directions and shifts in culture and identity of housing associations in the UK. Mapping key theories was critical to developing a framework for this study,

Table 1. Fieldwork outline

Research subject: A stand-alone BME subsidiary
Community outreach project
Strategic management
Operational management
Research methods
(a) Documentary analysis
Board reports, annual accounts
External funding and internal strategic reports
Organisational and outreach literature, e.g. pamphlets, leaflets, posters
(b) Semi-structured interviews: At 3–6 monthly intervals with key individuals
Outreach manager
Outreach case worker
Housing officer
Board Chairperson
Chief Executive
(c) Focus group
Outreach service users
(d) Participant observation and observation
Community events and workshops
Outreach steering committee meetings
Observing organisational behaviour

Source: The Author (2009).

particularly one which would seek to link logics to organisational culture as a theoretical outcome.

Institutions, or the 'omnibus conception' of institutions which Scott (2001) aptly refers to, are composed of

> ...cultural-cognitive, normative, and regulative elements that, together with associated activities and resources, provide stability and meaning to social life...institutions are multi-faceted, durable social structures, made up of symbolic elements, social activities and material resources. (Scott, 2001, p. 49)

Links can clearly be made between the intangible aspects of organisational culture (Johnson, 1992; Schein, 1997) and these 'symbolic elements' of institutions. Moving beyond the structure of institutions to their inherent qualities, Scott (2001) contends that although the functions of institutions are to provide stability and order, they themselves are subject to incremental and revolutionary change, and that institutions are thereby both a state and a process. Lounsbury (2007, p. 289) reiterates the association between the concept of logics and 'broader cultural beliefs and rules that structure cognition and guide decision making in a field'. Scott (2004) also refers to systems of logics within fields that range in terms of content, penetration, linkage and exclusiveness. Exclusiveness refers to what extent an institutional logic is dominant in its organisational field so that while some fields display somewhat consistent belief systems, other fields are typified by either secondary logics or 'multiple conflicting belief systems'. This view reinforces the earlier work by Friedland & Alford (1991) who strongly support the notion of multiple logics existing in all organisational fields, and specifically logics entrenched in a higher order of society which are hierarchical in form.

Applications of instutional logics to different industries were drawn upon to inform this study. This included Thornton's work (1999, 2002, 2004) on the book publishing industry which used historical analysis and extensive fieldwork to investigate shifts in logics over time particularly in relation to executive power and succession. It also included Mullins' (2006) sector level analyis of change in English housing associations which revealed 'a prolonged process of institutional adaptation has brought fundamental changes in patterns of accountability' (2006, p. 21). Considering key organisational change drivers and the emerging dynamics of the sector, he identifies the tension between pressures for hybrid housing associations to be both locally accountable, and efficient in the management and delivery of their social rental and development of social housing:

> Public policy drivers of efficient procurement and community engagement are in tension, and these tensions are played out in the dominant logics in the field and within individual organisations. (Mullins, 2006, p. 21)

In this case study of HAX, the focus shifts to the dynamics exemplified by one large and rapidly changing housing association. Strategy, culture and CI provided the key areas of enquiry for a multi-layered case study of a large organisation and acknowledged its co-existent realities as influenced by a particular set of internal and external change drivers, i.e. institutional logics.

Changes in CI at SHA

The key objectives of the Community service at SHA were:

> To ensure **equality** of access to culturally appropriate information, assistance and guidance in housing matters… To **create** a Specialist housing advice point for South Asian communities in need of housing and related matters. (SHA, 2008)

As described earlier, the CI outreach service at SHA was delivered through outreach advice surgeries, home visits, a confidential helpline and case workers attending appointments with clients. The termination of the external Lottery funding for SHA's CI project coincided with the restructuring programme at HAX, which was based on a rationale of centralising services to be more efficient and 'fit for purpose'. As the SHA outreach service was coming to a close, service users were invited to a focus group discussion, as part of this study by the author, to obtain their views on the value of the CI service, and which other agencies they would consider accessing after the closure of the service. Of the participants at the focus group, only one was a tenant of that association, while the others were members of the same local South Asian community and were in a variety of tenures, including city council housing, other housing association accommodation, private rented or living with extended family. All participants had ongoing housing cases with the SHA community programme, including serious repair issues or relocation applications.

The key themes that emerged from the focus group discussion were around cultural empathy, the personal service and a lack of an equivalent alternative housing advice.

The value of advice and advocacy in their own language was a strongly cited factor by participants who did not feel fluent or confident in English, and found the task of filling out forms and speaking to the local council or solicitors to be daunting. Their experience of mainstream local authority housing services was that: 'They don't speak in our language' (Service User, SHA, 2009).

Beyond the language barrier, service users felt that the community outreach staff, who were of a similar ethinic background, provided a culturally sensitive and empathetic service, allowing them to develop trust and build a relationship with their case worker. A few commented that the service had a good insight into which areas their families would feel comfortable and safe to live in. At the outreach service, staff made themselves available to make home visits, and attend meetings at banks or local authorities with service users. According to one participant in the focus group:

> [what is] Good about outreach service is able to come down your house and help you sort out your problems with full support. (Service User, SHA, 2009)

All the service users had contacted SHA with housing-specific concerns, and, despite guidance and signposting from the community project staff on which agencies in the future could further assist in their cases, there was a reluctance and unwillingness to consult more mainstream local authority services: 'They won't understand us—we don't want to go anywhere else' (Service User, SHA, 2009). Others described the closure of the service as 'upsetting' and 'very distressing' (Service User, SHA, 2009). When asked which agencies they would turn to after the service closure, local community groups were mentioned who also provided a general service targeted at the same South Asian community rather than any specific housing advice service.

Local agencies and community groups which had a working relationship with SHA's CI service were also contacted for feedback on their view of the programme. Overall the agencies and groups responded positively to working with SHA's outreach service, which they felt was valuable particularly because as it helped their clients overcome language and cultural barriers relating to accessing housing services.

In September 2008, a report was put forward to the SHA Board with proposals for the continuation of the outreach service after March 2009, when the external funding would expire. In highlighting the value of the community outreach project, the report made links between the service and some of the HAX's corporate objectives of 'Improving Lives', 'Increase Our Influence' and 'Provide Excellent Customer Services'. Furthermore, it was suggested that it was this community outreach programme that made SHA unique and established its important link with the South Asian community in that city. A number of options were considered for the future direction of the community project by trying to obtain funding to continue to run or expand the service through different grant streams or bodies. However, ultimately these options could not be realised because they were not feasible within SHA's business plan. The outreach project came to a close in June 2009, and a discussion was due to take place the following year as to the future direction of any CI programme at SHA.

Before the closure of its community service, a special housing services directory was produced by SHA to provide information to the South Asian community on other mainstream support providers and agencies to contact after the closure of SHA's outreach service.

Hybridity Enacted: Changing Organisational Culture and Institutional Logics

An important factor in considering the viable options for the future CI project was the funding climate in 2009, which had changed since SHA had initially secured the grant funding. In general, funders were not offering the large sums required to fund the entire project for a further period. In addition, funders were no longer offering grants for general housing advice services but concentrating on awarding grants for specific funding streams such as worklessness, health and well-being or financial inclusion. It was also becoming more challenging to obtain funding for BME specific projects, as funders were looking rather to award grants for neighbourhoods or areas rather than targeting specific minority groups. This shift away from localised and ethnically specific community services to specific themes of activities offered to a range of communities (Beider, 2010) was a trend that was reflected in and by changing political agendas, shifts in the nature of the delivery of CI among large social housing providers and indeed the strategic developments within HAX as an organisation.

Indeed, a year and a half prior to the closure of its community outreach programme, an erosion of SHA's links with the South Asian community had already begun with the loss of Asian staff and the association's offices moving from a predominantly Asian/ethnic minority area to a more 'professional' location in the business area of the city centre. Furthermore, as a result of the Choice-Based Letting system (a customer-focused approach emphasising tenant choice, introduced in England in the early 2000s), only a small percentage of SHA's few hundred properties were actually occupied by ethnic minority tenants, while only two of its Board members were from ethnic minority backgrounds. The link with the South Asian community had been an essential part of the BME identity of SHA, and this had been maintained through the CI service. With the termination of this service, the question was provoked as to what would now define the BME character of SHA, apart from its regulatory status.

Looking broadly at changes in organisational culture, in the early phases of the fieldwork, there was a widespread acknowledgement from board members, executives, housing managers and outreach staff at SHA of the value of the CI service as defining the character of the association, and making a unique contribution to a local community. This view of the outreach service as an asset to the organisation was evident in HAX's corporate literature. However, the strategic developments at HAX with its restructuring and centralisation programme driven by an efficiency logic did not seem to fit with the specialist local community service provided by SHA. Once external grant (and internal support) funding for the programme had expired, outreach staff felt that no central or head office resources were made available to support them in applying for further funding pools. Reflecting a growing business or customer-focused rationale, SHA was led down a more generic route with a standard housing service offer to its tenants, rather than a provider of specialist community outreach service.

In the space of few months, a cultural paradigm shift had occurred at SHA. Senior managers and Board members who were supportive of this highly individual community service where at the same time putting forward business arguments on 'where best to put limited resources' and advocating funding based on selective themes determined by the centralised CI department at HAX, rather than locally responding to the needs of a particular community. There were suggestions from senior staff that the community services offered by SHA were really the responsibility of the local city council. Furthermore, the 'loss of our outreach gives us some breathing space' (SHA Board Member, 2009) to think about the future direction of CI at the association and to concentrate rather on Tenant Involvement which would focus on SHA's tenants as customers specifically as opposed to a wider ethnic minority community. Board members with a personal and cultural link to the association, though rueing the loss of the outreach service, felt the financial stability achieved from being a subsidiary of an efficient and financially driven organisation such as HAX merited the shift in the identity of SHA. According to a Board member, since coming under the financial wing of HAX, SHA was becoming more professional, financially secure association with proper policies and procedures, and with valuable group central administrative support. However, there were also some perceived disadvantages of this relationship with the newly restructured HAX, such as greater delineated hierarchy, and a loss of self-determination and autonomy that led to management staff and the Board 'feeling a bit removed from the decision-making' (Board Member, SHA, 2009). With the positively perceived shift in organisational culture towards professionalism, there was also some apprehension that 'our organisation is in danger of losing its grass roots touch' (Community Service Manager, SHA, 2009).

Returning to the idea of multiple, sometimes competing, logics (Friedland & Alford, 1991, Scott, 2001; Thornton, 2004) and building on Mullins' (2006) application of institutional logics, the findings from this layer of the case study do reiterate the prevalence of a commercial, business logic which is tied in with a corporate and centralised culture, while a community logic has been more evident in the locally responsive culture of the terminated CI programme at SHA. The evidence presented here suggests that the dominant corporate sub-culture had largely overtaken the community-based, regional culture.

Building on this existing literature (Mullins, 2006), different sets of competing logics that best captured the empirical evidence from the overall case study of HAX were considered, before the author settled on 'Customer' and 'Community' logics as depicted in Figure 2. The logics/culture matrix in Figure 2 reflects the positioning of strategy according to the evidence presented in this layer and other layers of the case study, that is shifts in organisational culture and institutional logics through the strategic developments in CI services. Figure 2 provokes further questions regarding the strategic choices of the HAX as dominated by these logics and sub-cultures. In the empirical example of SHA, its previous community outreach programme could be clearly defined by a localist, community culture and logic that disappears or becomes homogenised as the structural changes are implemented across the organisation. To what extent the strategic direction of any future CI projects at SHA can be linked to the new, centralised vision of CI at HAX still remains to be seen.

Returning to the notion of hybridity enacted, this study has explored the tensions and conflicts of a third sector housing organisation whose strategies and culture have indeed been influenced by state, market and society forces as depicted by Gruis (2009). Drawing on Gruis's (2009) concept of contested influences on HA strategy, Mullins *et al.* (2009) considered four ways in which housing associations construct and prioritise their CI activities. This is illustrated in Table 2, where these four approaches are mapped as zones of hybridity.

Extending this line of argument, the author's logics–culture matrix in Figure 2 is further developed as an interpretive framework in Figure 3, to characterise CI at HAX and in the sector more broadly, adapting the arguments presented in the literature which are reinforced by the empirical evidence from this case study.

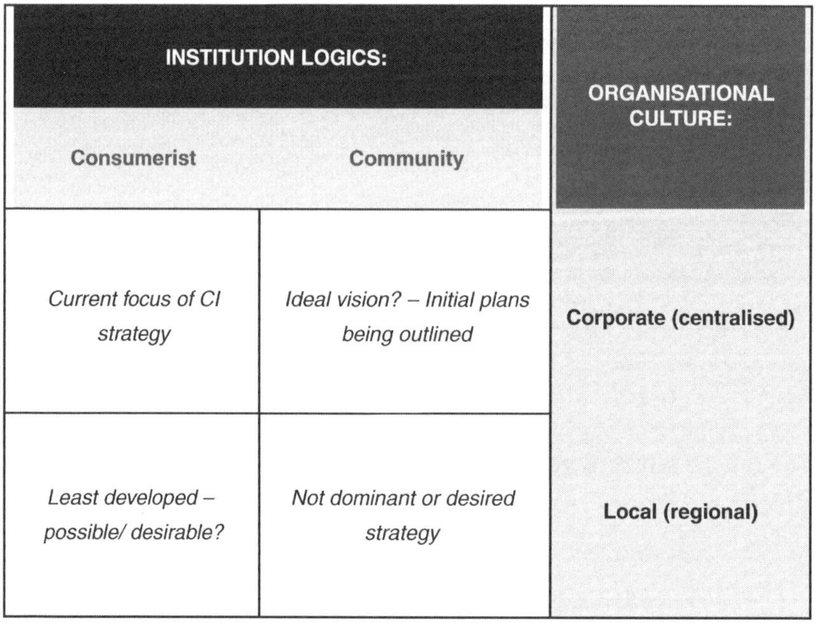

INSTITUTION LOGICS:		ORGANISATIONAL CULTURE:
Consumerist	**Community**	
Current focus of CI strategy	*Ideal vision? – Initial plans being outlined*	**Corporate (centralised)**
Least developed – possible/ desirable?	*Not dominant or desired strategy*	**Local (regional)**

Figure 2. A logics–culture matrix. *Source*: The Author (2011).

Table 2. Zones of hybridity: influences on CI activities

1. Strategy based	CSR becomes the corporate planning framework and priorities for social and CI activities are set and monitored corporately
2. Local relationship based	Priorities are set locally by local managers in partnership with residents and local community organisations
3. Contract based	Priorities are set externally by contracts won from state (and local state) who are seen as having the legitimacy to make these decisions
4. Partnership based	Priorities are negotiated externally through partnerships with other social actors (this is a strong theme for NHF and is evidenced in the NHF audit by the leverage of partner contributions)

Source: Mullins et al (2009)

The first approach is a 'Corporate Customer Approach' and is where, based on the empirical evidence, the author has located HAX. This type is closely linked to a strong market or 'strategy' approach as in Mullins *et al.*'s (2009) 'Zones of Hybridity' in Table 2.

At the time of this research study from 2007 to 2010, and in the Labour government years preceding it, a clear link is evident between a state/market approach (Mullins *et al.*, 2009) and a 'Network Community' approach as depicted in the second quadrant of Figure 3. At the time housing policy, supported by the NHF's 'In Business for Neighbourhoods' campaign promoted networks and partnership working in neighbourhoods through local strategic partnerships. In this network community approach, multiple communities are the collective beneficiaries of CI services.

The third approach in Figure 3 reflects a local customer focus. To some extent, this was being achieved by the operating companies at HAX prior to the centralisation of CI services. The approach suggests housing associations maintaining a business logic, with tenants still viewed as customers, but tailoring the CI product or service to meet their local need as private companies do with their customers. Medium-sized housing associations without the capacity, or indeed intent, to have a wider community impact could adopt this business model of CI.

The final type is the community specialist approach where a specific community is the service user or beneficiary. This approach is related to the 'local relationship' zone of hybridity as depicted in Table 2. A prime example of this was the outreach service at SHA, serving the specific needs of an ethnic minority community in a certain English city. This local community model is not the most economically feasible, which was the ultimate downfall of the specialist CI service at SHA. However for small, niche housing associations or indeed subsidiaries within larger HA groups, it could still perhaps provide unique, locally responsive community services as an added value to the organisation.

Conclusion

This paper has concentrated on one example within a larger case study to explore some of the shifts in priorities in the form of the strategic direction and sub-cultures of a housing

| INSTITUTIONAL LOGIC: | | ORGANISATIONAL CULTURE: |
CUSTOMER	COMMUNITY	
1. Corporate Customer Approach: • Market driven and business ethos • Tenants as individual customers are service users • Corporate Social Responsibility • Consistent or generic service offer • Large scale with big contracts • Scale of economies • Large housing associations with efficiency drivers	**2. Network Community Approach** • State/ market collaboration • Communities as collective entity are beneficiary of services • Partnerships and networks with local service providers • Neighbourhoods approach and community values • Large, medium and small housing associations working in partnership, leveraging scale and/or grass roots presence for maximum local impact	**Centrally driven** **Corporate business**
3. Local Customer Approach • Locally - responsive customer focus • Customer Tenants with specific local needs ascustomers • Tailored product or service • Possibly medium sizes housing associations without capacity or scale to have wider community impact, but still possible to tailor customer offer to respond to local needs.	**4. Local Community/ Community Specialist Approach** • Specialised or niche service offer responding to community needs • Specific community as service user/beneficiary • Responding to community needs • In current economic climate,this business model may only be feasible for small niche housing associations or subsidiaries with in larger housing associations providing unique service as added value.	**Locally responsive** **social activity**

Figure 3. Approaches to CI in the housing association sector. *Source*: The Author (2011).

association as related to its provision of community services over and above a core social housing provision. Relating this example to the general picture emerging from the analysis of primary evidence from across the organisation in the broader case study, the findings reiterate the tension between a more corporate, financially driven and centralist outlook which has dominated senior executive decision making and a more locally responsive, community-oriented, regional culture that has been increasingly displaced by the former. Both sets of sub-cultures seemed to be related to a number of historical and current external drivers relating to the identity and values of large housing associations. Implications of these emerging research findings for hybrid models are apparent including the question of how housing associations can balance broad, generic themes with specific, local community needs in the delivery of CI programmes. This question is pertinent in the context of limited resources and increasing pressures to do more with less, a trend heightened by impacts of the credit crunch and recession, and in all likelihood to be further reinforced by public sector cuts by the new coalition government in the UK (DCLG, 2010). With the increasing pressure on both private and public sector funding models, are highly personalised community support services targeting a specific section of society still financially feasible? Or are the valuable social benefits of specialised CI programmes for vulnerable communities a price worth paying? It is a debate that will no doubt continue. Choices between state, market and community drivers will become starker, and despite a rhetoric of localism, cost efficiency and centralisation are likely to drive both state and market influences on the sector. This does of course reinforce the hybrid identity of housing associations as driven by multiple institutional logics.

The multi-layered longitudinal design of this case study has provided a unique perspective from which to contribute to current debates about hybridity in housing. The concept of institutional logics has been built on and links made with organisational culture to explore the multiple realities and organisational sub-cultures tied in with institutional logics (Gregory, 1983; Schein, 1997; Scott, 2001; Thornton, 2004). This has more than repaid the decision to invest in a single multi-layered and longitudinal case study, and thereby provided glimpses of how a transition in institutional logics is enacted within a hybrid organisation.

One of the aims of the case study, on which this paper reflects, was to provide a unique contribution to existing knowledge in the field of housing research employing the concepts of organisational culture and institutional logics (Hofstede, 1983; Mullins, 2006; Schein, 1997; Scott, 2001). Of course, while the strategy adopted by the author was to employ more generalised, established conceptions of organisational culture, alternative theories of culture more specifically related to the third sector could be drawn on to further develop the links between culture and logics in understanding hybridity in social housing. No doubt further research will be needed to uncover the new strategic directions and shifting cultures both driving and driven by institutional logics and competing demands faced by social housing organisations in changing political and economic contexts. This future research could develop links between institutional logics and other fields of study that link organisational culture to more current public or third sector studies such as new public management (Hood, 2000; Hood & Peters, 2004) or even institutionalism in urban politics (Lowndes, 2009), particularly given the role of external change drivers in understanding the context of the changing course of CI in this case study.

Returning then to the big picture, in the first year of this study, sector debate was focused on political drivers, namely housing policy, regulation and impending legislation in the form of the Housing and Regeneration Act (July, 2008). In the second year of the research, political drivers gave way to the dramatic economic conditions of the international credit crunch and recession which had a significant impact on the strategic capabilities of housing associations. After the credit crunch, and with a new coalition government at the helm, it remains to be seen what the critical change drivers and institutional logics in the sector will be and how housing associations will respond to them, particularly the larger ones like HAX. The impacts of the new government agenda such as the UK Treasury's Comprehensive Spending Review in 2010 (setting out budgets for public sector departments to achieve deficit reduction targets) and the localism agenda of the coalition government (DCLG, 2010), which were beyond the scope of this research project, will no doubt provoke more debate on the changing role and character of housing association sector in England.

In the meanwhile, the multi-layered case study that is reflected on in this paper can hope to contribute to the application of institutional theory to research in the field of social housing, while providing further insight into the ever-changing dynamics of an intriguing sector and the identity of housing associations as hybrid organisations.

Acknowledgements

The case study project that was the subject of this paper is part of a 3 year collaborative research arrangement between the Centre for Urban and Regional Studies (CURS) at the University of Birmingham, and large English housing association, which has been anonymised for ethical reasons. This housing association, called HAX in this study, provided the author with an open access to all parts of its organisation and was supportive of the research process and the academic independence of the research throughout the duration of the study. The author is also grateful for the guidance, constructive critique and general encouragement from the PhD supervisors, based at CURS.

References

Beider, H. (2010) The declining significance of housing and race: Reviewing the housing corporation black and minority ethnic housing policy, Housing Studies Association Conference 2010, York.

Brandsen, T., Farnell, R. & Cardoso Ribeiro, T. (2006) *Housing Association Diversification in Europe: Profiles, Portfolios and Strategies* (Coventry: Rex Group).

Billis, D. (2010) Hybrid organisations and the third sector challenges for practice, *Theory and Policy* (Basingstoke: Palgrave).

Cave, M. (2007) *Every Tenant Matters: A Review of Social Housing Regulation* (Wetherby: Department of Communities and Local Government).

Clapham, D. & Evans, A. (1998) *From Exclusion to Inclusion* (London: Hastoe Housing Association).

Czischke, D., Gruis, V. & Mullins, D. (2010) Conceptualizing social enterprise in housing organisations, ENHR Conference, July 2010, Istanbul.

Department for Communities and Local Government (2007) *Cave Review of Social Housing Regulation: Responses to the Call for Evidence* (Wetherby: DCLG Publications).

Department for Communities and Local Government (2010) *Decentralisation and the Localism Bill: An essential guide* (London: DCLG).

Friedland, R. & Alford, R. R. (1991) Bringing Society Back, in: W. W. Powell & P. J. DiMaggio (Eds) *Symbols, Practices and Institutional Contradictions* (The New Institutionalism in Organisational Analysis, Chicago, IL: University of Chicago Press).

Glaser, B. & Strauss, A. (1967) *The Discovery of Grounded Theory* (Chicago, IL: Aldine).

Gregory, K. L. (1983) Native-view paradigms: Multiple cultures and culture conflicts in organizations, *Administrative Science Quarterly*, 28, pp. 359–376.

Gruis, V. (2009) *Conceptualising Social Enterprise in Housing* (Prague: Cecodhas Seminar), June.

Harrison, M. (1995) *Housing, 'Race', Social Policy and Empowerment* (Aldershot: Avebury).

Heino, J., Czischke, D. & Nikolova, M. (2007) *Managing Social Rental Housing in the European Union: Experiences and Innovative Approaches.* Final Report. CECODHAS European Social Housing Observatory and VVO-PLC (Helsinki: CECODHAS).

Hofstede, G. (1983) Culture's consequences: International differences in work-related values, *Administrative Science Quarterly*, 28(4), pp. 625–629.

Hofstede, G. (1993) Cultures and organizations: Software of the mind, *Administrative Science Quarterly*, 38(1), pp. 132–134.

Hofstede, G. (1998) Identifying organizational subcultures: An empirical approach, *Journal of Management Studies*, 35, pp. 1–12.

Hood, C. (2000) *The Art of the State* (Oxford: Oxford University Press).

Hood, C. & Peters, G. (2004) The middle aging of new public management: Into the age of paradox? *Journal of Public Administration Research and Theory*, 14(3), pp. 267–282.

Housing Corporation (2005) *Global Accounts of HAs* (London: Housing Corporation).

Johnson, G. (1992) Managing strategic change: Strategy, culture and action, *Long Range Planning*, 25(1), pp. 28–36.

Lounsbury, M. (2007) A tale of two cities: Competing logics and practice variation in the professionalizing of mutual funds, *Academy of Management Journal*, 50, pp. 289–307.

Lowndes, V. (2009) New institutionalism and urban politics, in: J. Davies & D. Imbroscio (Eds) *Theories of Urban Politics*, pp. 91–105 (London: Sage).

Malpass, P. (2000) *Housing Associations and Housing Policy: A Historical Perspective* (Basingstoke: Macmillan).

McDermont, M. (2010) *Governing Independence and Expertise: The Business of Housing Associations* (Oxford and Portland Oregon: Hart).

Mintzberg, H. & Waters, J. (1985) Of strategies, deliberate and emergent, *Strategic Management Journal*, 6(3), pp. 257–272.

Mullins, D. (1997) From regulatory capture to regulated competition, *Housing Studies*, 12(3), pp. 301–319.

Mullins, D. (2006) Competing institutional logics? Local accountability and scale and efficiency in an expanding non-profit housing sector, *Public Policy and Administration*, 21(3), pp. 6–21.

Mullins, D. & Pawson, H. (2010) HAs: Agents of policy or profits in disguise? Chapter 10, in: D. Billis (Ed.), *Hybrid Organisations and the Third Sector. Challenges for Practice, Theory and Policy* (Basingstoke: Palgrave).

Mullins, D. & Riseborough, M. (2000) *What are housing associations becoming? Final report of changing with the times project* (Birmingham, CURS: University of Birmingham).

Mullins, D. & Sacranie, H. (2008) Competing drivers of change in the regulation of HAs in England: A multi-layered merging perspective, in: *Housing Studies Conference*, York, 2–4 April 2008. Available at http://www.york.ac.uk/inst/chp/hsa/spring08/presentations/Mullins%20&%20Sacranie.pdf

Mullins, D., Latto, S., Hall, S. & Srbljanin, A. (2001) Mapping Diversity, Registered social landlords, diversity and regulation in the West Midlands Birmingham, Housing Research at CURS no 10.

Mullins, D. et al. (2008) Competing drivers of change in the regulation of housing associations in England: A multi-layered merging perspective, HSA Conference: York (unpublished).

Mullins, D. et al. (2009) Corporate social responsibility and the transformation of social housing organisations: Some puzzles and some new directions, ENHR Conference: Prague (unpublished).

Mullins (2010) HAs, TSRC. Working Paper 16 (University of Birmingham: TSRC). Available at http://www.tsrc.ac.uk/LinkClick.aspx?fileticket=qyS3AvWvt%2Bk%3D&tabid=500

National Affordable Homes Agency/Housing Corporation (2008) *Annual review of housing association private finance* (London: Housing Corporation).

National Housing Federation (2003) *In Business for Neighbourhoods* (London: National Housing Federation).

National Housing Federation (2008a) *The Scale and Scope of Housing Associations Activity Beyond Housing* (London: NHF).

National Housing Federation (2008b) Peers Second Reading Briefing 28 April 2008 p 2, 4.

Pawson, H. & Sosenko, F. (in press) The supply side modernisation of social housing in England: Analysing mechanics, trends and consequences, *Housing Studies*, forthcoming.

Schein, E. (1997) *Organisation Culture and Leadership*, 2nd ed. (Calif: Josey-Bass).

Scott, W. R. (2001) *Institutions and Organization*, 2nd ed. (London: Sage).

SHA (2008) Proposals for the Continuation of the Outreach Service post-31ST March 2009, *Report to the SHA Board*, September.

SHA (2008/09) *Community Outreach Service: Annual Report*.

Shwartz, H. & Davis, S. (1981) Matching Corporate Culture and Business Strategy, *Organisational Dynamics*, 10, pp. 30–48.

The Author (2008a) Corporate social responsibility in a socially responsible organisation: Corporate governance, culture and community investment in a case study of a large, not-for-profit English housing association, Unpublished PhD research paper. September 2008. University of Birmingham.

The Author (2008b) *Reflections on Responses to the Cave Review of Social Housing Regulation* (CURS: University of Birmingham), Available at: www.curs.bham.ac.uk.

The Author (2009) The impact of the credit crunch and economic slowdown on the dynamics of the UK social housing sector: A sector analysis with empirical evidence, Unpublished CURS 2nd year PhD workshop paper. February 2009. University of Birmingham.

The Author (2011) Strategy, Culture and Institutional Logics: A Multi-layered View of Community Investment at a Large Housing Association. Doctoral Thesis submitted to the University of Birmingham. Centre for Urban and Regional Studies: Birmingham.

Thornton, P. H. (2002) The rise of the corporation in a craft industry: Conflict and conformity in institutional logics, *Academy of Management Journal*, 45(1), pp. 81–101.

Thornton, P. H. (2004) *Markets from Culture: Institutional Logics and Organizational Decisions in Higher Education Publishing* (Stanford: Stanford University Press).

Thornton, P. H. & Ocasio, W. (1999) Institutional logics and the historical contingency of power in organizations: Executive succession in the higher education publishing industry, 1958-1990, *American Journal of Sociology*, 105(3), pp. 801–843.

Walker, R. M. (2000) The changing management of social housing: The impact of externalisation and managerialisation, *Housing Studies*, 15(2), pp. 191–202.

Webpages

Available at: http://www.hm-treasury.gov.uk/2010_june_budget.htm (accessed 25 June 2010).

Available at: http://www.cabinetoffice.gov.uk/sites/default/files/resources/building-big-society_0.pdf (accessed 15 March 2011).

Magical or Monstrous? Hybridity in Social Housing Governance

ANITA BLESSING

Amsterdam Institute for Social Science Research, University of Amsterdam, The Netherlands

ABSTRACT   *While a growing number of national social housing strategies rely on the work of hybrid entities blending social and commercial tasks, the state/market dualism continues to dominate the conceptual landscape of housing research. This exploratory paper develops a conceptual approach to support research into the role of not-for-profit social entrepreneurs in the housing market. It looks for insights within their 'hybrid' status, spanning state and market, and subject to multiple sets of institutional conditions. Four frames of hybrid identity are developed, and then substantiated via a discussion of two different sectors of not-for-profit social entrepreneurs in Australia and the Netherlands. As the growth trajectory of each sector is traced and the construction of hybrid identity is explored from both public and private perspectives, institutional pressures are revealed that set the current context for development. This brings forth implications for existing conceptual tools, as well as directions for new research.*

Introduction

> As long as housing is seen as 'either state or market', we will not be able to analyze the complex interrelation between state and market mechanisms in real-world housing. (Bengtsson, 1995, p. 2)

Housing researchers face an ongoing challenge rooted in the nature of housing itself. Its status in the modern welfare state context as *both* 'an individual market commodity' *and* 'a public good demanding state involvement' (Bengtsson, 1995, p. 2) presents a complex institutional reality. To explore this complexity, researchers construct, recycle and *de*construct concepts in order to contain data and build theories (Sartori, 1970). The process is ongoing, with shifting empirics prompting new models of thinking (Kemeny, 1995 p. xiv).

A core concept in housing research is the dualism of 'state and market' or 'public and private sectors'. Its power is far-reaching, with 'state' and 'market' recognised 'as necessary analytical notions without which it is impossible to understand or describe modem life' (Salamon & Anheier, 1992, p. 126). Within a housing context, this dualism supports differentiation between public and private actors and institutions. Citizen ownership, governance via elected officials and bureaucratic distribution, for example, contrast markedly with shareholder ownership, corporate governance and market mechanisms. From this process of differentiation, ideal types emerge, framing public perceptions. The state/market dualism also supports analysis of relationships between housing actors. For example, Bengtsson (1995, p. 4) grounds housing 'policy theories' in different configurations of state and market actors, including *state only, state complemented by market, state corrected by market, market only* and *market complemented by state*. Observing that housing's role as a market good favours pursuit of social goals via market mechanisms, he posits a *dominant* policy theory in modern welfare states of *market corrected by state*.

While the state/market or public/private dualism retains currency, its explanatory power is limited in the face of change. In line with Bengstsson's claim, national housing strategies rely increasingly on market corrections via private not-for-profit entities that transcend sector boundaries. They originate from varying national patterns of development. Two main patterns include subsidised mass growth of not-for-profit housing companies exemplified by the Netherlands and Sweden (see Priemus, 2008; Turner, 2007), and gradual transfers of funding or dwelling stock from government agencies to the not-for-profit sector, observable in England, the USA and Australia (see Bratt, 2009; Malpass & Victory, 2010; Milligan *et al.*, 2009). Common to all these settings has been the more recent institutionalisation of not-for-profit housing providers as social entrepreneurs. Rather than relying on state subsidies, these organisations use limited state support to lever private development capital and to pursue commercial profits for social ends (see Mullins & Pawson in Billis, 2010).

The rise of these social entrepreneurs constitutes a significant change that remains academically under-explored. Efforts to understand this trend require new conceptual tools to sharpen the research focus. Changes also prompt re-evaluation of existing tools, such as theoretical models of national rental housing markets. Do they retain sensitivity to changing empirics?

This exploratory contribution aims to expand the conceptual basis for research into the rise of not-for-profit social entrepreneurs in the housing market. It departs from the notion that these organisations have many faces. Not-for-profits are alternately 'scorned for evading the laws of the market-place', and celebrated as 'the locus of values, voluntarism, pluralism, altruism (and), participation' (DiMaggio & Anheier, 1990, p. 153). Similarly, 'like the elephant in the famous story of the blind men, different insights about the concept of social enterprise emerge as it is viewed from different perspectives' (Cordes & Steuerle, 2009, p. 5). Insights into the identity of these organisations are thus sought within their 'hybrid' status, spanning state and market, combining public and private action logics, and subject to multiple sets of institutional conditions.

In this paper, four frames of hybrid identity are developed to guide research into the role of not-for-profit social entrepreneurs in the housing market. A discussion following two different sectors of these organisations, the small Australian sector and the mature Dutch sector, then provides an opportunity to substantiate the conceptual approach. As the growth trajectory of each sector is traced and the construction of hybrid identity is explored from both public and private perspectives, institutional pressures are revealed that set the current context for development. This brings forth directions for research and provides an opportunity to review existing theoretical tools.

The choice of empirical examples for discussion follows the logic that insights into the role of hybrid organisations arise as borders and boundaries are crossed. As organisations operating across state and market, not-for-profit social entrepreneurs epitomise hybridity (Brandsen *et al.*, 2005a) (Billis, 2010, p. 15). By exploring sectors of these organisations within Australian and Dutch rental markets, the discussion encompasses extremes of hybrid identity. Characteristically placed at opposing poles in theoretical models, both markets are undergoing dynamic, border-crossing changes. Within what Kemeny would term a 'dualist' rental market, the embryonic Australian sector is being groomed for mainstream market activity, including moderate-income housing provision. In a 'unitary' rental system (see Kemeny, 1995), the mature Dutch sector is now called upon to curb its entrepreneurial activities and narrow its client base. In line with the aim of guiding new research, the contrasting cases hold value as a differentiated basis of evidence (see Oxley, 2002, p. 89).

The first part of this paper lays out the conceptual tools used in the approach. It discusses existing tools, clarifies key terms, considers criteria for effective conceptualisation and develops a new approach. The second part applies this approach to an empirical discussion. A concluding section draws out theoretical implications and identifies directions for further research.

Developing Hybrid Identity for Housing Research

Existing Conceptual Tools: Theories of the Development of Social Housing Markets

While international transfer of housing policies warrants understanding of the contextual implications involved, deep and systematic comparisons are rare (Oxley, 2002). To address this problem of 'transferability', or to support comparisons, researchers draw on the existing literature. Studies by Harloe (1995), Kemeny (1995, 2001), Kemeny *et al.* (2005), which model different rental markets, are frequently cited in relation to contemporary cases. Harloe differentiates between social housing consumption in the form of the *mass* model, characterised by broad access, and the *residual* model, which limits access to those deemed 'most in need' (Harloe, 1995). Kemeny differentiates instead between policy structuring of cost and profit forms of rental (Kemeny, 1995, p. 178). In his dualist model, government co-opts and residualises cost rental, sectioning it off from the market, limiting it to those in high need, and promoting homeownership. His *unitary* model, by contrast, denotes competition between cost and profit rental. A final model, the *integrated* rental market, is based on harmonisation of cost and profit rents.

The models by Kemeny and Harloe are drawn upon to support classification of international social housing systems (see for example, Czischke, 2009). They also help frame debates regarding convergence or divergence of international housing policy contexts (see Gruis & Priemus, 2008). Despite superficial similarities, the causal assumptions underpinning the models differ significantly. Harloe maintains that housing is best seen as a key commodity and economic driver, rather than a pillar of the welfare state (Harloe, 1995, p. 3). Social housing must supplement, not supplant the for-profit market. Thus, the mass model of social housing, which poses 'a challenge to core capitalist interests' (p. 538) is best seen as a transitional economic strategy and not as a sustainable model of social welfare. On this basis, Harloe posits international convergence towards the residual model. In direct contrast, Kemeny comes at housing from a welfare state perspective, going so far as to argue, 'it is time housing began to take its rightful place at the centre of welfare research' (Kemeny, 2001, p. 68). He explores potential for state housing policy to strategically determine rental market dynamics. Stressing the potential for divergence of international social housing markets, he advocates pursuit of the 'integrated' rental market (Kemeny *et al.*, 2005, p. 870).

While the two differing conceptual models continue to support comparative housing research, recent changes in social housing provision prompt re-evaluation of their currency. To what extent do they capture the dynamics of contemporary provision? Malpass & Victory, for example, argue that Harloe's models fail to explain the continued remodelling of social housing, including processes such as privatisation (Malpass & Victory, 2010, p. 6). This contribution both draws on models by Kemeny and Harloe, and considers some aspects of their applicability to the present state of play, with a focus on the privatisation and marketisation of social housing.

Defining 'Not-for-Profits' and 'Social Entrepreneurs'

'Not-for-profit social entrepreneurs', an awkward composite term that helps to describe modern social housing provision, is here unpacked and defined. In recent years, the term 'not-for-profit' has gained popularity over 'non-profit', in recognition of a growing reliance on profit-making activities within the sector. Salamon & Anheier complain of a 'terminological tangle' in defining the 'non-profit' sector, with terms such as 'charitable', 'independent', 'voluntary', 'tax-exempt', and 'non-governmental organisation' (NGO) emphasising certain attributes of the label and clouding its meaning (Salamon & Anheier, 1992, p. 128).

A general set of formal and informal conditions attached to private 'not-for-profit status' is used in this discussion, with losses in 'connotative precision' compensated by gains in 'extensional coverage' (Sartori, 1970, p. 1035). First and foremost, not-for-profit status is acquired on the basis of a formally instituted social mission, be it charitable or more broadly prescribed. Further, a legal constraint placed on the distribution of profits to owners is in place to promote adherence to this mission. There may also be constraints on commercial ventures unrelated to the social mission. In return, not-for-profits receive state support such as tax-concessions, subsidies, cheap credit or low-cost land. Hand-in-hand with this support comes public accountability, instituted via both formal regulatory requirements and informal societal expectations.

An entrepreneur 'sets up a business or businesses, taking on financial risks in the hope of profit' (*Concise Oxford Dictionary, 2001*). Modified by the term 'social' it becomes a catchphrase returning 8 700 000 hits on Google, yet it lacks a widely accepted formal definition. A common reading of 'social entrepreneur' is an entity 'using the disciplines of the corporate world to tackle daunting social problems' (Eakin, 2003). This is consistent with the notion of a social enterprise channelling commercial profits to fund social projects in the field of housing.

Criteria for Concept Development

Sartori describes concepts as both data containers and as the building blocks of theories (Sartori, 1970). What then, are the attributes that enable concepts to fulfil these functions? Before the concept of hybridity is developed as an approach to the research focus, some basic criteria for effective concept development are considered.

Following Deutsch, Salamon & Anheier (1992, p. 136) cite *economy, significance* and *predictive powers* as criteria for concept evaluation. 'Economy' refers to the ability to simplify and order reality while staying true to its form. 'Significance' signals the ability to highlight attributes relevant to current debates, so that a new conceptual construct reverberates with others around it. 'Predictive powers', a more complex criterion, is not about predicting causal pathways. Rather, it is achieved by concepts that bring forth hypotheses, lending momentum to new research. A further indicator of this predictive quality is 'rigour', or travelling power for future use. Hybridity as a conceptual approach to the research problem should thus be able to capture its defining elements, highlight crucial issues arising from it, bring forth ideas for new research, and be applicable to new cases.

Hybridity as a Conceptual Approach to Social Entrepreneurs in the Housing Market

Brandsen *et al.* (2005a p. 6) define hybridity within a governance context in terms of organisations and ways of working that cross-cut 'state, market and civil society', blending ideal types, cultures, co-ordination mechanisms or action logics. Along with a number of other studies, they explore the concept of hybridity as an approach to not-for-profit, or 'third sector' identity (see also Billis, 2010; Czischke *et al.*, 2010; Mullins & Pawson in Billis, 2010). From these studies emerges a dilemma. Is hybrid identity best defined loosely to reflect ambiguity and fluidity? Or should researchers pursue a tighter definition, classifying sub-types and mapping their attributes?

Highlighting the risks of 'stumbling into hybridity' (Billis, 2010, p. 17), Billis makes the case for a more rigorous definition. Rather than relying on simplistic interpretations such as the mere association of hybridity with the blurring of boundaries, he argues that 'practice, policy and theory now have an urgent need for a tougher conceptual approach to the phenomenon' (p. 3). He presents a model of nine hybrid zones overlapping public, private and civil sectors, between which organisations may rapidly shift (pp. 57–58). While they stress the interpretive nature of their findings, Czischke *et al.* (2010) make a similar attempt to build conceptual rigour by gleaning components from existing third sector classifications to fit the housing context.

Brandsen *et al.* display less confidence in the potential for researchers to pin down the identity of 'third sector organisations'. To illustrate the challenges involved, they draw on the notion of hybridity, while acknowledging the ambiguity and fuzziness that surrounds it. Arguing that hybrid organisations are 'difficult to identify on the basis of traditional concepts', they advise that we should 'accept and understand them as they are, not in terms of static ideal types' (Brandsen *et al.*, 2005a, p. 760). They illustrate this idea with the example of the Griffin, calling it a 'fantastic creature that can only be described in terms of its constituent parts, which by implication means that it has no clearly defined identity of its own' (p. 759). Turning instead to the chameleon, which adapts to changing surroundings whilst retaining a distinct identity, they position hybridity and change as core to third sector identity. Hybrids, they argue, might best be classified 'by their strategies, as methods of adaptation to conflicting demands' (p. 759).

The conceptual approach developed below aims to strike middle ground between classificatory power on the one hand, and sensitivity to change on the other. It rests on two assumptions. First, counter to the idea that 'hybridity can be described only in terms of the elements it is built from' (Brandsen *et al.*, 2005b, p. 1), it repositions hybridity as a locus of strong and distinct meanings. Viewed thus, the Griffin is more than part eagle, part lion.

By blending qualities of the king of the air and the king of the land, it spans Heaven and Earth, signifying divine power (Eason, 2008, p. 83). Thus, its meaning (A + B = C) far transcends the sum of its component parts (A + B = AB). A second assumption is that hybrid identity shifts according to the perspective taken, with different readings competing for dominance.

Four Symbolic Interpretations of Hybridity

Over time, the concept of hybridity has amassed layers of meaning through expansive use. Characteristically used in biology, the label 'hybrid' is applied to an increasing range of heterogeneous products, systems and entities, from cars to musical genres to forms of governance. Hybrid creatures feature as prominently in contemporary popular culture as they do in ancient mythology. The approach taken here draws these different layers of meaning together with the rhetoric surrounding the role of not-for-profit social enterprises in the housing market. It identifies a series of ideal-types used to frame hybrid identity in different ways. These ideal-types may symbolise strategies, mark empirical shifts, or help to decode existing processes of political framing.

Hybridity as a state of transformation. The frame of *transformation* echoes the emphasis by Brandsen *et al.* (2005a, p. 749) on perpetual adaptation as a hybrid trait. In a governance context, it signals the blurring of sectoral boundaries as state/market configurations shift, along with funds and responsibilities. The frame of transformation is grounded in mythology, wherein shape-shifters signify both dualisms and the transitions that bridge them. Just as a mermaid shape-shifts to cope with the transition between water and land (Bell, 1991, p. 322), a not-for-profit may turn entrepreneurial in the face of funding cutbacks. Brandsen *et al.* explore this frame through Ovid's transformation of Arachne into a spider. Noting that she retains her identity as a weaver—and thus her

Figure 1.

integrity, despite change, they suggest that this same ability to evolve whilst keeping defining qualities intact may offer a key to understanding third sector organisations

While transformation may be a valid lens through which to view the hybrid form, it leaves much to be explained. It does not account for stable hybrid arrangements, nor does it reveal the causes of change. To return to the example of Arachne, her transformation into a spider was neither voluntary, nor a coping strategy. Rather, she was bound to a new form as a punishment for pride. Thus read, the frame of transformation suggests transgression and reform. Clearly, a satisfactory explanation for continuing organisational change must run deeper than the notion of shifting funds and responsibilities. What other facets of hybrid identity might help explain these processes of change?

Hybrids as links between cultures.   From the foreign origins of the Sphinx and Chimera, hybrid creatures of Ancient Greek mythology (see Burr, 1993, p. 273), comes a second frame of hybridity. It positions hybrids as cross-cultural icons or *links between cultures*. The not-for-profit sector thus becomes an agent of cross-pollination and institutional mediation, helping government to connect with local networks, enlisting for-profit partners and laying the cultural foundations for ongoing relationships. 'It is possible that in bouncing between different environments, hybrid organisations may serve to transfer elements between those environments' (Brandsen *et al.*, 2005a, p. 8).

The frame of links between cultures also manifests within the socio-cultural 'grassroots' value attached to the not-for-profit sector. Billis (2010, p. 10) taps into this dynamic, attributing the public appeal of hybrid organisations to their ability to mobilise communities around a problem, identify needs, offer 'joined-up services' and promote volunteering. Yet alongside this positivity lurks the ever-present question as to how hybrid organisations may develop and adapt (like Brandsen *et al.'s* chameleon) without losing touch with community roots. Does grassroots appeal limit potential for growth?

Figure 2.

Figure 3.

Hybrid vigour. The concept of *hybrid vigour* or heterosis, which describes the inherent strength of hybrid organisms, makes a compelling frame of hybrid identity underpinned by stories of powerful and magical creatures like the Griffin and the Centaur (see Burr, 1993, p. 60). The prospect of obtaining a better or stronger individual by combining the virtues of its parents relates well to a governance context, with hybrid organisations often idealised as super-blends of ethical drive, professional acumen and practical know-how. 'Hybrid vigour' permeates the rhetoric surrounding the minimisation of the state and the resulting shift towards 'governance' of social housing outcomes delivered via the not-for-profit sector (see Rhodes, 1996, pp. 653 and 667). Here, the hybrid form takes on a distinct, even magical identity, with innovative new partnerships based on legal and economic advantages, leading to 'institutional complementarity'.

As evidenced in biology, with some crossbreeds proving to be weak or infertile, hybrids are by no means inherently powerful. The cross-sectoral action characteristic of hybrid organisations carries multiple risks, revealing 'vulnerability' as the flipside to hybrid vigour. Hybrid organisations must straddle different sets of rules, draw on diverse sources of support, and invest in multiple forms of accountability.

Hybridity as a transgression. A further aspect of hybridity is rooted in the etymology of the term. Derived from the Latin word 'hybridia' (itself from the Greek *hubris*), hybridity denotes 'the offspring of a tame sow and wild boar'—two progenitors amusingly reminiscent of the state and the market. The term 'hybrid' may thus be interpreted as 'an outrageous miscegenation' (*The Concise Oxford Dictionary*, 2001 p. 695). In a competitive economic context premised upon the dichotomisation of market and state, the flexibility of hybrid status, which blends 'action logics' and combines social and commercial functions, may be read as duplicitous, unjust or threatening. Thus viewed, hybridity takes on monstrous qualities, with hybrid organisations positioned as *transgressions* of the binary opposition of state and market.

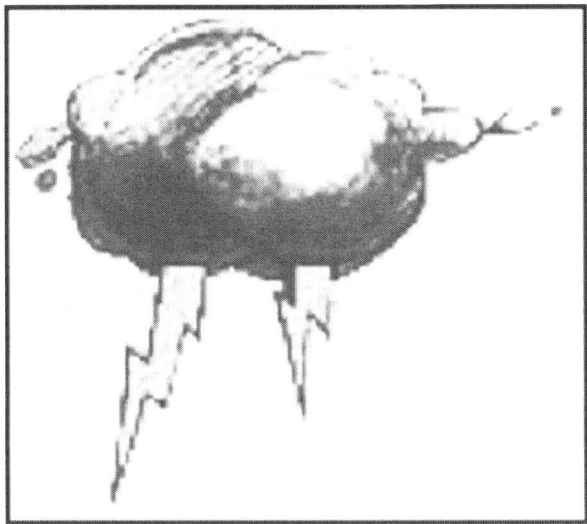

Figure 4.

The frame of transgression also resonates from a public accountability perspective. The recent British government culling of quangos (quasi non-governmental organisations) provides an example, with government decrying its predecessor for 'putting important decisions into the hands of unelected quangocrats' (The Conservative Party, 2010).

Having developed these four frames of hybrid identity, the task remains to put them to work within a housing context and give body to this conceptual approach. This is done through a discussion of not-for-profits institutionalised as social entrepreneurs in Australian and Dutch housing markets. By tracing the growth of each sector and exploring the construction of hybrid identity from both public and private perspectives, the discussion highlights the institutional pressures that set the current context for development.

Australian and Dutch Social Housing Provision: The Growth of Not-for-Profits as Social Enterprises and the Construction of Hybrid Identity

Comparing Social Housing in Australia and the Netherlands

In the decades following the Second World War, 90 per cent of Australian adults passed through homeownership (Beer *et al.*, 2007, p. 13). While homeownership remains the dominant tenure, at 68–70 per cent of households, a 400 per cent rise in house prices over the last 20 years has pushed it out of the reach of many. Households in the private rental sector (21–25 per cent) suffer low tenure security. An estimated shortage of 493 000 dwellings that are both affordable and available to low-income tenants has led to tough conditions for renters, which are heightened in urban areas. Social housing stands at just 5 per cent of dwellings (National Shelter, 2010). In the very different housing policy context of the Netherlands, housing problems are similar. A general shortage of housing that meets contemporary needs is amplified in certain locations. House prices have tripled over recent decades, and despite policies favouring homeownership (now 57 per cent of households), it is slipping out of the reach of many. In both the for-profit private rental sector

(11 per cent of households) and the large social housing sector (32 per cent) demand is high, leading to illegal social sub-markets, soaring rents in 'free-sector' dwellings and severe shortages in urban areas (Ministerie van Volkshuisvesting, Ruimtelijke Ordening en Milieubeheer, 2009, p. 151).

While contrasting tenure patterns in Dutch and Australian rental housing do not make for easy comparisons, there is common ground worth considering. In line with the policy theory of 'state correctives to the market', private not-for-profits active in the housing market as social entrepreneurs are positioned within both countries as the preferred providers of social housing, with a key role in boosting housing supply. With both sectors in transformation, the Australian and Dutch examples reveal issues arising at opposite poles of Kemeny's continuum of rental markets. In 2009, as the Australian Federal Government rolled out reforms to encourage not-for-profits into the market, the Minister of Housing described a need for more 'large, commercially sophisticated not-for-profit housing organizations' possessing 'the flexibility and commerciality we need to transform our social housing system' (Plibersek, 2009). That same year, when the Dutch Minister of Housing proposed a suite of reforms, he took a different tone. The large and sophisticated Dutch social-entrepreneurial sector had revealed an 'Achilles heel' in internal control. A better balance was needed between public accountability on the one hand, and autonomous social entrepreneurship on the other (Van der Laan, 2009). These contrasting readings of hybrid identity prompt examination of the trend towards social entrepreneurship. What is driving it and how does it unfold within different policy contexts?

Australia

Large-scale public housing provision began in Australia prior to the Second World War as a means of spatially aligning labour and industry. Despite a surge in provision during and after the war, sometimes referred to as the 'golden era' of Australian public housing (Hayward, 1996, p. 29), private housing investment was eventually prioritised and public housing remained a minor tenure (Beer *et al.*, 2007, p. 13). Significant sales of public housing stock occurring from the 1950s have led to the characterisation of Australian authorities as 'reluctant landlords', unwilling to engage in large-scale development except in instances where public housing is required to support industrial growth (Hayward, 1996). During recent decades, other 'measures to suppress public renting' (Kemeny, 1995, p. 111), such as surplus-generating rents and use of surpluses to pay housing allowances, have accompanied residualisation of the sector. This process culminated in 2005, when the state of New South Wales ended secure tenure for new tenants and made renewal of leases subject to demonstration of continuing need. While it remains a crucial source of support, Australian public housing is now formally entrenched as a 'landlord of last resort' in a policy context consistent with Kemeny's 'dualist' model. However, when the 'dualist' label is applied to a broader sector now repackaged as 'social housing', does it still fit? In 2009, when the Nation Building Economic Stimulus Program provided funding for new public dwellings, construction occurred under the proviso that stock would soon be transferred to the not-for-profit sector. These new assets would lever private development capital and support entrepreneurial projects aiming to deliver 'affordable' social housing to medium income households.

The growth of the not-for-profit sector. The Australian not-for-profit (community) housing sector has grassroots origins, formalised by top-down government programs from the late 1970s (Bisset & Milligan, 2004 p. 12). The Community Housing Federation of Australia now represents over 2000 organisations across the country. Varying estimates place their rapidly growing portfolio at less than 100 000 dwellings (see National Shelter, 2010). Top-down initiatives contributing to the sector's growth have framed hybrid identity as 'links between cultures', emphasising mediation and cross-pollinating powers. A 1993 government-commissioned study reasons that non-profits 'being less formidable (than government) and more in touch with community and assistance agencies, should fill a gap' (Mant, 1993, p. 1). It takes note of a further advantage of the community sector in its ability to connect government to new sources of funding. An Australian academic study taps into this rhetoric: 'Increasingly non-profit community services are centrally implicated in mediating the type and quality of relationship between the state and its citizen/subject' (McDonald & Marsten, 2002, p. 377).

In the 1990s, the Federal Government positioned the not-for-profit housing sector as 'a true competitor for public housing' and required the states to plan for its growth. In a manner consistent with Bengtsson's policy theory of 'state corrected by market' (Bengtsson, 1995), the term 'social housing' crept into the discourse to blur the boundaries between public and community housing responses, minimising political risks associated with state administration of a market good. When a new Federal Government redirected funding to private rental subsidies, state authorities shifted stock to the community sector, where tenants could access this significant stream of financial support. While not-for-profits were growing, they owned only half of the stock they managed, and remained dependent on government grants and concessions linked to Charitable or Public Benefit Institution (PBI) status under Australian Taxation Law (CFHA, 2009).

Instituting social entrepreneurship. From the beginning the 21st century, supply-side intervention through the not-for-profit sector was favoured. Housing affordability had emerged as a problem affecting a wide range of households. 'Affordable' rental housing for moderate-income households and key workers (on a smaller scale than in Harloe's 'mass' model) gained political currency. But how could supply-side assistance for these mainstream groups be delivered? Within the context of an ideological shift in public service provision towards 'governance' of privately provided services, state housing authorities announced accelerated programs of stock transfers to the not-for-profit sector. While the rate of transfers actually occurring failed to match the intensity of the rhetoric around this new mode of privatisation, the approach formed the cornerstone of growth strategies. Planned expansion of the sector would come from providers' own assets and borrowings, with government investment providing a lever. So began the government push for not-for-profit social entrepreneurship in the housing market.

To further this approach a new frame of hybrid identity was needed. The catchphrase *'community sector partnerships operating commercially to deliver public value'* captures this period's growing reliance on the frame of 'hybrid vigour', wherein arrangements blending state and market action logics lead to better ways of meeting mainstream housing needs. In the 2007 lead-up to a Federal election, the Howard Government Housing Minister announced a sudden policy shift. Public housing funding was to be reallocated via competitive tenders to applicants across sectors with proposals for affordable housing.

Evoking 'hybrid vigour' in the form of a magical relationship between cross-sectoral action, innovation and rapid outcomes, he declared:

> Let's open it up to innovation, let's see over the next two months what industry can come up with, what the NGOs can come up with, what councils and the states can come up with – I think we'll be pleasantly surprised as to how much innovation can be achieved, how much leverage that can provide ... and at the end of the day, that means families with roofs over their heads (Fraser & Maley, 2007).

Following a change of government, the new Minister of Housing placed the same emphasis on hybridity, synergy and rapid results. Leaning on the frames of 'links between cultures' and 'hybrid vigour', she outlined a 'vision for the way forward', with not-for-profits as the glue within partnerships that 'play to the strengths of the respective players' (Plibersek, 2008). Describing 'large, commercially sophisticated not-for-profit housing organisations' as 'the centrepiece of the Government's reform agenda', she expressed hope that more would emerge, 'operating in different markets—including across State borders—providing a range of housing products for low and moderate income Australians' (Plibersek, 2009). As the director of one of the fastest growing community housing organisations explained:

> We build housing just like as any other builder and developer would, but rather than passing profits through to shareholders, we can retain those and plough them back into producing more housing (Werden, 2008).

A key program for growth of the sector is the National Rental Affordability Scheme (NRAS), launched in 2008. Via annual incentive payments, the NRAS sets out to bring together for-profit and not-for-profit investors and developers to achieve rapid, large-scale outcomes in the form of 50 000 new dwellings by 2012, and a further 50 000 dwellings subject to demand. These must be rented to eligible tenants at 20 per cent below market rates for a 10-year period. The scheme is itself a hybrid creation that takes considerable risks with international policy transfer. It resembles a 'lite' version of the US Low-income Housing Tax Credit (LIHTC), reliant on capital gain, and lacking long-term use requirements. In the absence of regulations compelling banks to invest locally, it faces a challenge in securing funds. Unlike the LIHTC, the NRAS will be administered by a centrally regulated not-for-profit sector (as in the British model) with ex-public assets shouldering market risks. It will target a broad client group, more reminiscent of Dutch or Swedish social housing systems.

While the public sector construction of hybrid identity leans heavily on the frame of 'hybrid vigour', concerns raised by non-government stakeholders in the NRAS highlight vulnerability as the flipside of its hybrid identity. Perceived points of weakness are often attributed to 'competing institutional logics' (see Mullins, 2006). In line with the 'links between cultures' frame, community housing associations concerned that rapid growth could compromise local connections evoke the familiar conflict between 'scale and efficiency and local accountability' (Mullins, 2006, p. 6). In the words of one industry stakeholder: 'We don't want to become a commercial animal' (Stakeholder interview, 2008). Academic and housing industry studies stress the not-for-profit sector's lack of development experience (Milligan et al., 2009).

From a private for-profit perspective, hybrid identity in the Australian housing market takes on a different face. A Property Council of Australia submission on the NRAS emphasised the gap between state and market housing agendas, and protested the inclusion of low-income households, 'it is inappropriate to expect institutional investors to take on high risk public housing tenants' (Property Council of Australia, 2008, p. 12). A conflict between the social need to house low-income people and the market need to secure investment had become apparent. More serious objections came from an economic legal perspective. Several months into the implementation of the NRAS, the Australian Tax Office issued a sudden warning. Non-profit participants already up and running with projects faced loss of their charitable status on the grounds that moderate-income housing provision transgressed the requirement of limiting activities to poverty relief. In pushing for social-entrepreneurship in housing provision, government had overlooked this part of the deal. The hybrid NRAS scheme had tripped on its own institutional foundations.

In response, the Federal Government passed a 'temporary legislative extension to charitable purpose' allowing community housing organisations to continue their NRAS projects. Further correspondence from the Tax Office questioning whether the social housing classification 'low-income' qualified a household as 'poor' enough to receive charitable assistance made a longer-term solution seem unlikely (CHFA, 2009). Yet a recent turn of events offers hope. In a ruling unconnected to the field of social housing, the High Court held that any commercial activities carried out by a charity would be considered charitable as long as profits were directed to its charitable mission (Australian Federal Treasurer, 2010). It seems the definition of charity has now been stretched to encompass the primary requirement of social entrepreneurship.

Institutional challenges. In the current context of Australian housing development, hybrid identity is cast in a positive light. Mass media celebrate the rapid growth of not-for-profit social entrepreneurs, proclaiming 'community housing is big business' (Werden, 2007). Yet beneath this stream of positivity, considerable divergence between public and private perspectives on hybrid identity gives cause for concern. While moderate-income housing has long been a small part of the business of community housing organisations, the government quest for hybrid vigour through the large-scale NRAS has brought this side of their work under the spotlight as a transgression. With the requirements of Kemeny's dualist rental market written into taxation law, the decision to pursue models of provision normally found in a 'unitary' policy context at first seemed fraught with risk.

Despite the surprise ruling allowing charities to carry out commercial activities, not-for-profit participation in the mainstream market is still at a small scale. If realised, the government's vision of large, commercially sophisticated social entrepreneurs is likely to require other forms of compromise between social and commercial interests in the housing market. Social norms, such as the mantra of assisting those in 'highest need' may continue to enforce the state /market dualism. With social housing tenants stigmatised as a financial risk, finding the investment partners to realise the vision may be difficult.

The Netherlands

The Dutch constitution sets out 'the promotion of adequate housing opportunities' as 'the subject of government care' (Article 22, Clause 2). True to this provision, government has generally acted as a promoter, rather than as a direct provider of social housing; yet for substantial periods of time it has done so in a manner that responds to mainstream needs. In the Netherlands, Kemeny *et al.* (2005) find an example of a unitary rental market, defined by a lack of 'barriers preventing non-profits from competing with for profits' and an ingrained acceptance of the social market philosophy. While the presence of cost rental does influence rents, dualisms in the rental market are apparent, with properties outside the social market attracting high rents due to under-supply.

The growth of the not-for-profit sector. Dutch social housing emerged from a pillarised system of political and religious groups that provided housing, labour unions and schools outside structures of government (Salet, 1999). Harloe identifies multiple sources of support for a private alternative to state housing arising as the need for social services grew at the turn of the 19th century. Resistance to state intervention in housing and education on the part of religious factions was one source. Efforts by workers' organisations to take control of their own housing also provided support. Top-down charitable initiatives by a bourgeoisie needing to preserve class structures lent further momentum to the private approach (Harloe, 1995, pp. 25–29). An analysis of early social housing tenure conditions concluded: 'supervision, control, education and discipline were the basic concepts of the residential civilising process of the Dutch working-classes' (Deben, 1989, p. 289). This drive to preserve existing social structures of class, trade and religion constructed hybrid identity in a manner that best fits the frame of 'links between cultures'. The frame of transgression was also discernable in early efforts on the part of commercial entrepreneurs to protect their housing market share and keep state agencies out of the game. This provided a third source of support for a private alternative to state housing (Harloe, 1995, pp. 25–29).

The institutionalisation of housing associations via the Housing Act of 1901 began a hybrid system wherein financial and legal privileges were allocated by government in exchange for the fulfilment of social tasks. These arrangements expanded to a new scale of action when a severe post-war housing shortage prompted government to engage the sector in the implementation of a large centrally planned construction agenda. This hybrid relationship with housing associations facilitated large-scale housing development to meet mainstream demand. Despite their huge market share (a case of 'state complemented by market'), housing associations had not yet become social entrepreneurs, and functioned more like government branch offices (AEDES, 2007). In the decades that followed, state support bound them to local or regional operations where they negotiated their own rules in the relative absence of commercial stimuli (Salet, 1999). During this time, the sector professionalised and developed the infrastructural foundations of its later role. The 'Social Housing Guarantee Fund' (WSW) was established to support asset management, along with the Central Fund for Housing (CVF), a redistributive fund for housing associations in financial stress. By the mid-1980s, not-for-profits were the primary producers of housing (Kemeny *et al.*, 2005, p. 869).

Instituting social entrepreneurship. The transition to social entrepreneurship in the Dutch context occurred in the mid-1990s, in a policy context characterised by 'government austerity, deregulation, market conformity, privatization and promotion of home-ownership' (Priemus, 1996, p. 1981). Central government, burdened by rising operating subsidies, negotiated an exchange of the value of outstanding government loans for future property subsidy commitments (Priemus, 2008). In the years that followed, housing associations benefited from low interest-rates, rising rent-levels and soaring property values, emerging into this century as a network of asset-rich social entrepreneurs. They soon diversified their activities in the property market, advantageously combining the freedom of entrepreneurial risk-taking with the security of collective resources backed by state guarantees. In 1998, the CVF expanded its role to include financial supervision of the industry.

As owners of social housing and other assets, the approximately 450 Dutch housing associations that are active today have enjoyed a strong financial position since the reforms of the mid-1990s. Despite the fact that corporation tax exemptions have ended (Priemus, 2008), access to low-cost land and credit and other financial concessions continues to support the provision of key social services, including large urban renewal projects beyond the reach of government. With their hybrid form and collective infrastructure grounded in corporatist links, their scale a product of 'mass' social housing provision and their role as social entrepreneurs opening up opportunities within a competitive market, contemporary Dutch housing associations embody the ideal of hybrid vigour.

While the Dutch social housing sector possesses all of the qualities Australian government strategies aim to cultivate, the grass on the other side of the fence is not always greener. Dutch housing associations' control of massive social assets has become controversial from a public accountability perspective. Their failure to meet housing needs has earned them the label 'sleeping giants' (see Berkelder *et al.*, 2003, p. 1), suggesting that growth has corrupted their community-based mission. Negative perceptions have led to media scrutiny over what they do and how they do it. High executive salaries, Maserati-driving CEOs and golden handshakes have made headlines (see Temminga, 2006; Vermeer, 2009). Risky commercial investment practices have been framed as representative of a hybrid identity synonymous with transgression (Stakeholder interview, 2008). A recently proposed policy response clips housing associations' entrepreneurial wings, limiting the proportion of investment capital they may put into any commercial project to 33 per cent, with the rest to be provided by for-profits (Van der Laan, 2009).

Heightening these internal pressures for reform is a formidable external pressure from the European Commission. As the former EU Commissioner explains:

> If we think of the European economy as a football match, I set and enforce the rules of the game. We make sure it is a fair match and that there is punishment for people and companies that break the rules and spoil the game for others. (Kroes, 2009)

As recipients of state-support, Dutch housing associations are viewed as having an unfair advantage in market competition, with formal complaints by Dutch for-profit investors adding momentum to the frame of transgression. Following a 2009 agreement between Dutch Central Government and the EU, housing associations are required to leave mainstream market activities to their for-profit peers and focus their efforts on households earning less than €33 000 (CECODHAS, 2010).

Institutional challenges. While Dutch housing associations continue to be valued as providers of core social services, the negative framing of hybrid identity from both public accountability and private market perspectives sets the context for future development. Having been framed as a transgression, the Dutch social housing sector is set to undergo a transformation.

Two main institutional challenges are apparent. The first lies within EU Competition Policy. In the words of one industry stakeholder, the EU 'thinks in terms of either market or public domain. Anything in between is suspect' (Stakeholder interview, 2008). A 2005 letter from the European Union to the Dutch Government states 'letting homes to households that are not socially deprived cannot be regarded as a public service' (Dormal Marino, 2005) and advises sale of any stock surplus to this core task. Having developed in a manner consistent with Kemeny's unitary model, the Dutch social housing system now has to adapt to dualist institutional requirements. This has reinforced the construction of hybrid identity as a transgression of the state/market divide. Changes to the targeting of social housing set to take effect early in 2011 could take a toll. As the hybrid is pushed out of the mainstream housing market, it is estimated that half a million households rendered ineligible for social housing will now be caught between a welfare system they cannot access and a commercial market they cannot compete in (CECODHAS, 2010).

A second institutional challenge lies in restoring a sense of legitimacy to the hybrid form as a social service provider, following transgressions on the public accountability front. One industry stakeholder explained: 'In our case, it's very difficult to define to whom we are accountable' (Stakeholder interview, 2008). Having transcended both their community roots and their financial dependence on government, housing associations are private entities with core social responsibilities that bear significant private risks. Despite this, 'if you ask the man on the street, he will say they are public' (Stakeholder interview, 2008). While commercial autonomy must be balanced with public accountability, taking market risks does not always sit well with prudent stewardship of public assets.

Conclusions

This exploratory paper developed a conceptual approach to hybrid identity for research into the role of not-for-profit social entrepreneurs in the housing market. Through a discussion of the rise of these organisations within two different policy contexts, it examined the construction of hybrid identity from both public and private perspectives. This revealed institutional challenges shaping the context for development in each national setting, and provided an opportunity to review existing conceptual tools the light of contemporary empirics.

Kemeny's and Harloe's models of rental housing markets both confront an issue core to contemporary social housing provision: how the balance is struck between social and commercial interests in the housing market. From a private market perspective, Harloe's findings help explain the challenges of negotiating terms for the market participation of social entrepreneurs. Kemeny's welfare state perspective highlights the potential for state policy to shape housing markets in divergent ways, achieving the 'hybrid vigour' of harmony between social and commercial interests, or patrolling the borders between them.

Both perspectives reveal limitations. Reinforcing Malpass & Victory's observation, Harloe's models of consumption fail to account for the dynamics of privatisation and marketisation (see Malpass & Victory, 2010, p. 10). Tensions in the Australian context between the dual objectives of accommodating low-income households and attracting private investment in social projects reveal that extending services to include some middle-income households has become a financial imperative for social housing providers. The migration of social housing provision into the private market thus prompts a revision of the residual and mass models. Kemeny's models show a similar lack of sensitivity to privatisation. Since Kemeny made the claim that the advantage of electoral accountability has led to a preference over the 'grass-roots model' for 'state-owned and state-managed cost renting ... as a means of providing public rental housing that the private market has been unable to provide'(Kemeny, 1995, p. 99), the tables have turned in favour of *privately owned and privately managed affordable renting as a means of providing social rental housing that the state has been unable to provide*. This may have implications for the feasibility of complex strategies such as the 'integrated market', which require strong powers of institutional design on the part of the state.

In light of these limitations, what can be gained from an approach that gives consideration to both state and market perspectives? Application of the four frames of hybrid identity revealed contrasting developmental challenges. The Australian Government is publicly courting hybrid vigour as a magical solution to the problems of a severely dichotomised housing market. A small community-based housing sector is positioned for rapid growth as a provider of low and moderate-income housing. Within a dualist policy context, institutional foundations for this more integrated approach now need to be developed. In the Netherlands, a mature not-for-profit housing sector is framed as a monstrous transgression from both public interest and private market perspectives. While a unique path of development has equipped the sector with powers that embody the ideal of hybrid vigour, it is now rendered highly vulnerable. Policy shifts in favour of a dualist model threaten to recast hybrid identity as a state of liminality, with hybrids relegated to the threshhold between the domains of state and market.

These developmental challenges show that social entrepreneurship is not a super-blend, but a balancing act. The different, and sometimes incompatible sets of rules that apply to social and commercial pursuits require trade-offs to be made, and this means compromise. Grounds must be negotiated for the private market participation of hybrid entities in receipt of state support. Public accountability for their work must be instituted without diminishing their capacity for entrepreneurship. Based on this perspective, further comparative research will examine how this compromise between social and commercial goals is being negotiated within different types of rental housing markets. As dualist rental markets embrace social entrepreneurship, and unitary markets adapt to dualistic requirements, the institutional foundations for these changes, both formal and informal, will be examined. Inclusion of cases in between Australian and Dutch extremes will provide for deeper analysis and evaluation of hybridity in social housing governance. At a time when supranational economic requirements are being enforced, and international social housing policy transfer is on the rise, research of this nature will build understanding of the institutional conditions needed to support different policy approaches over the long term.

References

AEDES (2007) *Dutch Social Housing in a Nutshell*, May (Hilversum).

Australian Federal Treasurer (2010) Letter from Treasurer Wayne Swan to Carol Croce, the Community Housing Federation of Australia, 14 January.

Beer, A., Kearins, B. & Pieters, H. (2007) Housing affordability and planning in Australia: the challenge of policy under neo-liberalism, *Housing Studies*, 22 (1), pp. 11–24.

Bell, R. (1991) *Women of Classical Mythology: A Biographical Dictionary* (Oxford: Oxford University Press).

Bengtsson, B. (1995) *Housing—Market Commodity of the Welfare State* (Gävle, Sweden: Uppsala University, Institute for Housing Research).

Berkelder, M., Fit, J., Smeding, H., Vreeker, J., Hoek, S. & Schoon, B. (2003) *Property Research: Nederlandse Woningcorporaties- Slapende Reus Wordt Wakker* (Research Department, Kempen & Co: Amsterdam, The Netherlands).

Billis, D. (Ed.) (2010) *Hybrid Organizations and the Third Sector: Challenges for Practice, Theory and Policy* (Basingstoke: Palgrave Macmillan).

Bisset, H. & Milligan, V. (2004) *Risk Management in Community Housing*. Report for the National Community Housing Forum, Sydney.

Brandsen, T., van de Donk, W. & Putters, K. (2005a) Griffins or chameleons? hybridity as a permanent and inevitable characteristic of the third sector, *International Journal of Public Administration*, 28(9), pp. 749–765.

Brandsen, T., Ribeiro, T., van Hout, E. & Putters, K. (2005b) Hybridity: a distinct identity, Third Sector Study Group Conference of the European Group of Public Administration, Bern.

Bratt, R. G. (2009) Challenges for non-profit housing organizations created by the private housing market, *Journal of Urban Affairs*, 31(1), pp. 67–96.

Burr, E. (1993) *The Chiron Dictionary of Greek and Roman Mythology* (Chiron Publications).

CECODHAS Housing Europe (2010) Half a million households in the Netherlands excluded? Available at http://www.housingeurope.eu/

CFHA (2009) *Submission to the Henry Review of Australia's Future Tax System*.

The Concise Oxford Dictionary (2001) (Oxford: Oxford University Press).

The Conservative Party (2010) *Quango plans to enhance accountability*. Available at http://www.conservatives.com/News/News_stories/2010/10/Quango_plans_to_enhance_accountability.aspx

Cordes, J. & Steuerle, C. (2009) *Non-Profits and Business* (Washington DC: The Urban Institute Press).

Czischke, D. (2009) Managing social rental housing in the EU: a comparative study, *International Journal of Housing Policy*, 9(2), pp. 121–151.

Czischke, D., Gruis, V. & Mullins, D. (2010) Conceptualizing social enterprise in housing organisations. Paper presented at ENHR—Urban Dynamics & Housing Change, Istanbul, July.

Deben, L. (1989) Residential civilization in The Netherlands 1850–1969: the rules and regulations stipulated in leases, *Journal of Housing and the Built Environment*, 4(3), pp. 289–302.

DiMaggio, P. & Anheier, H. (1990) The sociology of nonprofit organizations and sectors, *Annual Review of Sociology*, 16(1990), pp. 137–159.

Dormal Marino, L. (2005) Letter regarding the EC's provisional judgement on the financing of Dutch housing associations (Brussels: Europese Commissie, Directoraat-Generaal Concurrentie).

Eakin, E. (2003) How to save the world? Treat it like a business, *The New York Times*, 20 December.

Eason, C. (2008) *Fabulous Creatures, Mythical Monsters, and Animal Power Symbols* (Greenwood Press: Westport, CT, USA).

Fraser & Maley (2007) Howard's housing takeover, *The Australian*, July 27. Available at: http://www.theaustralian.com.au/national-affairs/climate/howards-housing-takeover/story-e6frg6xf-1111114048961

Gruis, V. & Priemus, H. (2008) European competition policy and national housing policies: international implications of the Dutch case, *Housing Studies*, 23(3), pp. 485–505.

Harloe, M. (1995) *The People's Home? Social Rented Housing in Europe and America* (Oxford: Blackwell).

Hayward, D. (1996) The reluctant landlords? A history of public housing in Australia, *Urban Policy and Research*, 14(1), pp. 5–35.

Kemeny, J. (1995) *From Public Housing to the Social Market* (London: Routledge).

Kemeny, J. (2001) Comparative housing and welfare: theorising the relationship, *Journal of Housing and the Built Environment*, 16(1), pp. 53–70.

Kemeny, J., Kersloot, J. & Thalmann, P. (2005) Non-profit housing influencing, leading and dominating the unitary rental market: three case studies, *Housing Studies*, 20(6), pp. 855–872.

Kroes, N. (2009) Neelie Kroes, European Commissioner for Competition: 'Policy Private Enforcement of State Aid Rules', State Aid Conference Brussels, 19 October.

Malpass, P. & Victory, C. (2010) The modernisation of social housing in England, *International Journal of Housing Policy*, 10(1), pp. 3–18.

Mant, J. (1993) The future of community tenancy schemes. Phillips Fox Solicitors, 10 November. Available at http://johnmant.com/gallery/pdf_125422026462154.pdf

McDonald, C. & Marston, G. (2002) Patterns of governance: the curious case of non-profit community services in Australia, *Social Policy & Administration*, 36(4), pp. 376–391.

Milligan, V., Gurran, N., Lawson, J., Phibbs, P. & Phillips, R. (2009) *Innovation in affordable housing in Australia: Bringing policy and practice for not-for-profit housing organisations together*. Final Report No. 134, AHURI, Melbourne.

Ministerie van Volkshuisvesting, Ruimtelijke Ordening en Milieubeheer (2009) *Cijfers over Wonen, Wijken en Integratie* (*Figures for Housing, Community and Integration*).

Mullins, D. (2006) Competing institutional logics? Local accountability and scale and efficiency in an expanding non-profit housing sector, *Public Policy and Administration*, 21(3), 6–24.

Mullins, D. & Pawson, H. (2010) Housing associations: agents of policy or profits in disguise, in: D. Billis (Ed.) *Hybrid Organizations and the Third Sector: Challenges for Practice, Theory and Policy* (Basingstoke: Palgrave Macmillan).

National Shelter, Inc. (2010) *Housing Australia Factsheet* (National Shelter, Inc).

Oxley, M. (2002) Meaning, science, context and confusion in comparative housing research, *Journal of Housing and the Built Environment*, 16(1), pp. 89–106.

Plibersek, T. (2008) Speech to the NSW Community Housing Conference.

Plibersek, T. (2009) Room for More, Speech, Sydney Institute, 19 March.

Priemus, H. (1996) Recent changes in the social rented sector in The Netherlands, *Urban Studies*, 33(10), pp. 1891–1908.

Priemus, H. (2008) Real estate investors and housing associations: a level playing field? The Dutch case, *European Journal of Housing Policy*, 8(1), pp. 81–96.

Property Council of Australia (2008) Submission regarding the NRAS Queensland Community Housing Coalition (QCHC) (2008) Submission to the Senate Inquiry into the NRAS Bill.

Rhodes, R. A. W. (1996) The new governance: governing without government, *Political Studies*, 64(3), pp. 652–667.

Salamon, L. & Anheier, H. (1992) In search of the nonprofit sector I: the question of definitions, *Voluntas*, 3(2), pp. 125–161.

Salet, W. (1999) Regime shifts in Dutch housing policy, *Housing Studies*, 14(4), 547–557.

Sartori, G. (1970) Concept misformation in comparative politics, *The American Political Science Review*, 64(4), pp. 1033–1053.

Stakeholder interviews (2008) Carried out by the author in Amsterdam (September, 2008) and Sydney (August, 2008).

Temminga, M. (2006) Gouden handdruk overstijgt miljoen (Golden handshake exceeds one million), *NRC Handelsblad*, 11 July 2006.

Turner, B. (2007) Social housing in Sweden, in: Whitehead, C. & Scanlon, K. (Eds) *Social Housing in Europe* (London: London School of Economics and Political Science).

Van der Laan, E. (2009) Brief Minister Van der Laan inzake voorstellen woningcorporaties, telsel, 12 June, 2009.

Vermeer, O. Een Toezichthouder die blaft, maar ook kan bijten (An overseer who barks, but can also bite) *NRC Handelsblad*, 11 June 2009.

Werden, C. (2008) *Community housing a big business*, ABC, Inside Business. Available at http://www.abc.net.au/insidebusiness/content/2007/s2427283.htm

Index

accountability 156, 157, 164–5; housing associations in England 131, 132, 133, 135, 137, 138; non-for-profit housing organisations in Australia 87, 160; public housing in US 60, 67; social-entrepreneurial sector in Netherlands 158, 163, 164
Aiken, M. 21, 111, 115
Alcock, P. 18
Alter, K. 113
Amin, A. 111, 113
Anheier, H. 2, 3
Arachne 154–5
Arbib, M. 86
Aspire 125
Australia 3, 6, 15, 150; National Rental Affordability Scheme (NRAS) 73, 74, 79, 80, 81, 85, 86–7, 160, 161; New Public Management 43, 74; not-for-profit housing organisations in Australia *see separate entry*
Austria 25

Baron, R. 58
Beer, A. 157, 158
behaviour variables 5, 21, 25, 27–9, 30, 55; *see also* programmatic hybridity
Beider, H. 140
Bell, R. 154
Bengtsson, B. 149, 150, 159
Berkelder, M. 163
Berndt, H.E. 45
Bidet, E. 16
Billis, D. 1, 2, 3, 5, 15, 21, 23, 25, 35, 55, 56, 61, 76, 80, 81, 84, 87, 111, 129, 131, 151, 153, 155
Bisset, H. 76, 85, 159
Blackburn, R. 113
Blessing, A. 2, 3, 74, 75, 84, 85, 87
Boelhouwer, P. 15
Bornstein, D. 111
Brandsen, T. 2, 5, 16, 21, 25, 27, 30, 38, 74, 75, 94, 100, 106, 133, 151, 153, 154–5

Bratt, R.G. 36, 37, 40, 41–2, 43, 44, 45, 47, 48, 49, 50, 150
Brennan, G. 79
Brouard, F. 15, 17
Buckingham, H. 2, 5, 15, 21
Burr, E. 155, 156

Carmel, E. 7, 112
Carter, M. 40, 47
Castells, M. 94
Cave, M. 130, 133
centralisation: housing associations in England 130, 135, 136, 139, 141, 143, 145
charitable status 10, 159, 161
Charlotte Housing Authority (CHA) 54, 60–1, 68; HOPE VI 54, 57–9, 60, 61–5, 66, 67–8; Moving to Work (MTW) 54, 59–60, 65–6, 67, 68; non-profit and for-profit subsidiaries 55, 60–1, 65; public–private–third hybrid zone 55; tax credits and for-profit corporations 63; tenant selection 67
China 7, 106, 107; Hong Kong 94, 106
Chiu, R.L.H. 94, 95
Chua, B.H. 94
Cisneros, H. 58
Clapham, D. 133
classification framework for social enterprise in housing organisations 24–9
classification models for social enterprise 19–24, 38
co-operatives 111
Cohen, R. 45
community investment (CI) and social welfare type services: employment *see separate entry*; housing associations in England *see separate entry*; Korean social services 100; US social services 48
competition, low wage 28
competition policy 10, 163, 164
conceptual approach: four frames of hybrid identity 149–51, 153–4; Australia 150, 151, 157, 158–61, 164, 165; conclusions 164–5; criteria for concept development 152–3;